My Experiments with Project and Personnel Management

My Experiments with Project and Personnel Management

SEKAR KUMBESWARA

PARTRIDGE

A Penguin Random House Company

ISBN: Hardcover   978-1-4828-5907-2
             Softcover   978-1-4828-5908-9
             eBook   978-1-4828-5906-5

Print information available on the last page.

To order additional copies of this book, contact
Partridge India
000 800 10062 62
orders.india@partridgepublishing.com

www.partridgepublishing.com/india

Table of Contents

About the Author

Mr. Sekar is the sixth born of Mr.Kumbeswara Iyer and Saraswathi, born at a town named Kumbakonam in Thanjavur District of Tamil Nadu in India in a family of ten children. He had his basic education in Kumbakonam, initially in Native Primary School, then in Town High School and later did his pre-university course in Government Arts College on the other banks of Cauvery river, fondly known as the 'Cambridge of South India'.

He then completed his Bachelors of Engineering in Dr. Alagappa Chettiar College of Engineering and Technology (ACCET), Karaikadi under Madurai University in 1975. He went on to pursue his Masters of Technology at the prestigious Indian Institute of Technology, Madras (I.I.T.) specializing in Transportation Engineering.

He was appointed as Lecturer in Civil Engineering department in Siddhaganga Institute of Technology, Tumkur, in Karnataka in 1978.

Selected by Union Public Service Commission (UPSC) of India, he joined as Assistant Engineer in Andaman Lakshadweep Harbour Works (ALHW), Ministry of Shipping, Government of India at Port Blair on 26th December 1978, exactly 26 years before the fateful tsunami hit the emerald islands.

Mr. Sekar subsequently rose through levels as Executive Engineer, Deputy Chief Engineer and finally as Chief Engineer and Administrator which is the highest position within the organization. He was valued with extreme kindness in his organization, administration and the Ministry of Shipping as a person with profound subject matter expertise and extraordinary caliber unseen in that wing of Central government services.

During his tenure as Executive Engineer, Mr. Sekar was nominated by the Government of India to pursue a course at The International Institute of Hydraulic and Environmental Engineering under Delft University of Netherlands in 1987 with a focus on Ports and Harbour Engineering. During this period, he had the opportunity of studying the latest technology and functioning of major international ports like Amsterdam, Rotterdam, Le Havre, Belgium, Germany, France etc.

During his service in Andaman Lakshadweep Harbour Works, his contributions to various important projects are being remembered across not just his own organization but across various other public sector organizations even after his relocation outside the islands following his superannuation in September 2012. The multiple projects he has envisioned and delivered in the form of dry dock, break waters, jetties, wharves, vehicle ferry ramps, bridges, prestigious buildings, roads, water supply systems etc. stand testament not only to his experience in the field of Engineering but also to his expertise in people management and managing upwards.

His contribution in the tsunami rehabilitation work was phenomenal and his ability in Project Management was greatly appreciated by one and all in the islands. He also served as Chairman of Institution of Engineers (India), Port Blair chapter for two terms. Mr. Sekar brought 52 Central Government organizations together as the Chairman of Central Government Employees Welfare Coordination Committee (CGEWCC) for a period of more than seven years where he created perfect harmony and coordination among all these organizations. Apart from his academic background and his professional accomplishments, Mr. Sekar is a great orator and his staff attended the meetings presided by him specially to hear his speech rich with interesting anecdotes. He is a patient listener and truly compassionate towards people's problems. He always strived to alleviate their challenges to the fullest of his abilities. His tenure in the department is still valued as the golden period by the staff and workers.

Acknowledgements

This is the greatest occasion for me to express my gratefulness indebted to many people for having shaped me to what I am today and made me stronger and successful in my endeavors.

Firstly, I thank Prof. Siddhalingappa, Head of the Department of Civil Engineering in Siddaganga Institute of Technology, Tumkur, Karnataka who was a pillar of support to me when I was working at the institution in 1978 as Lecturer (Civil). But for him, I could not have swiftly transitioned from my then current position to Andaman islands to join Andaman Lakshadweep Harbour Works (ALHW).

My heartfelt gratefulness to M/S.G.Raghavan, M.Gnanaolivu, R.Sivaswamy, W.S.J.Thambudurai and Punnoose Cherian Ex-CE&As, ALHW who have greatly influenced me and shaped me to what I became towards the end of my tenure. Also, I have Mr. N. Kannan (Retd. Additional Chief Engineer, ALHW) to thank for his contribution in exposing me to the dynamics of navigating complex and tricky situations.

I have fond memories of working with Mr.B. V. Balasubrahmanyam (Retd. Deputy Chief Engineer, ALHW). It was from him I learnt the art of approaching any problem with an open mind and the art of effective communication, to write and speak.

I take this opportunity to thank all my colleagues, officers, junior engineers, staff and workers who stood by me in our "ALHW Family"- people who were always ready to respond to my call at any hour of emergency and stood with me to ensure success in each and every endeavor of my journey through projects.

I am extremely grateful to those close friends who swam by my side when the "tide was the lowest" and virtually lifted me from sinking, offering me the unwavering courage to rise up and stand firmly.

Last but not the least; I fondly remember my own family. First of all, my dear and loving wife who has been so helpful and supporting while I was running around every island to attend to the tsunami rehabilitation works in Andaman and Nicobar Islands after 26ᵗʰ,December,2004. She recognized my role and stood by me and my team during the rehabilitation efforts. Now, with this book published, I will have the satisfaction of acknowledging her contributions in making my family and my career most successful.

My son Jai, daughter (in-law) Anjali, daughter Jyothsna and son (in-law) Vijay were with me throughout my journey of writing this book and supportive with their inputs.

At the beginning of each heading, you will find a quote which is very selective and close to my heart. They may not be relevant to the heading under which they are standing but are surely thought provoking for humanity. Some of them have the name of authors and some do not. You may see a few of them with author's name as "LSS". Don't be surprised if I say that LSS means L. Subbulakshmi Sekar who is none other than my wife. Many people used to come to her to get a reprieve for their grievances and she advices them with lots of patience. These quotes have stood the test of time and are based on valuable life experiences. I offer a special thanks to my dear wife for her contributions to this book even without her knowledge.

Last but not the least, I record my gratitude to all those authors and publishers who have kindly granted me permission to use their quotes and materials which are very much relevant to my chapters and which are useful to my readers.

I pray the Almighty of God to bless all these people stated above and also everyone in this Universe with happiness, peace, prosperity, wealth, health and long life.

K.Sekar

Preface

It was early summer in Chicago, United States of America, when I landed with my wife Mrs. Subbulakshmi Sekar in the June of 2014 to relax and spend our time in the company of our grandson Aarohan all the more than with our son Jai Anand and daughter (in-law) Anjali. Most importantly, we were there to attend the prestigious convocation of my son's MBA at Kellogg Graduate School of Management under Northwestern University in addition to Post Graduate Diploma in Business Management (P.G.D.B.M) from Management Development Institute (MDI), Gurgaon. We were also joined by my daughter Jyothsna and my son (in-law) Vijay who flew from India exclusively for this event. They also hold degrees in management studies from prestigious institutions of India.

From the Regents Park Apartments in Hyde Park Boulevard where we stayed, the view of Chicago was breath taking! Seeing through the glass walls of the apartment from twelfth floor we enjoyed the beautiful beautifully mowed green lawns of the Harold Washington Park. Cars on the road below the building seemed to move with a robotic obedience, hardly honking and always stopping at the traffic lights patiently even late at night! It all felt too good to be true for me and my wife who are used to the unforgiving Indian traffic system where 'traffic rules are considered suggestions'.

We had very long and extended evenings as the sun never sets here before eight in the evening.

The most beautiful sight from our apartment was the vast spreading Michigan Lake beyond the park with huge spread of water, waves splashing sometimes on the banks when it is windy giving the feel of a sea. Most notably, this lake stores potable water. How I wish we had more of such huge lakes in India, at least one in each state. I always felt strongly that nothing is impossible in engineering! My dream of such huge fresh water reserves in each state of India can be a reality if there is a strong political will to interconnect the main rivers of India. It is high time that a comprehensive master plan is drafted with minimum compromise towards environment so that the human lives are saved in India. But then, I digress.

What seemed to us like an unpredictably changing climate was always forecast accurately by the Meteorological department of Chicago that pushed notifications to my kin's handheld devices. For the first time in my life, the attire for the day was decided by the numbers on my son's mobile phone.

The trip was getting exciting!

Coming back to our family get together at Chicago, my children persuaded me to talk about various projects I had executed during my career as a Specialist in Ports and Harbour Engineering and to discuss the best and worst moments during the course of my service. That set the ball rolling! I started narrating all such interesting incidents and how I handled all such situations of emergency. Though it all started very casually, soon it picked up a different momentum and dimension! The application of mind in those incidents narrated lead to a serious discussion on current management techniques.

My son, daughter, daughter-in-law and son-in-law – all being management graduates from top tier business schools were surprised to note that I have been practicing most of their current day management techniques throughout my service.

They insisted on my going into action on my long time desire that I start penning down my experiences in the form of a book which would be of great inspiration to budding managers. This has been my long time desire which was ignited more after our sharing.

In this book, I have attempted to link my hands-on experience of 32+ years to the latest management techniques which would be useful not only for aspiring managers and engineers but also will benefit those aiming to reach senior level executive and CEOs positions.

Foreword

I deem it an honour that Mr K. Sekar asked me to write the foreword of his book 'My Experiments with Project & Personnel Management.' I was fortunate in seeing him in action as the Chief Engineer and Administrator of the Andaman Lakshadweep Harbour Works, in fact he was my comrade in arms as went about not just rebuilding the Port assets like jetties that were destroyed by the terrible Tsunami of December 2004, but of creating new infrastructure in the islands. I am told that even as I write, islands like Katchal, Chowra and Theresa will be connected to the outside world directly when their jetties that can accommodate inter-island ships are inaugurated. Mr K Sekar may have retired but these were all planned and started from scratch when he was at the helm. He is a hands-on leader whose technical knowledge and wide experience enabled his department to deliver extraordinary results in the face of difficult circumstances.

The book is a reflection of his career from his early days till he reached the pinnacle. But it is not so much about him as about how to draw valuable lessons from different experiences. And because each is a learning experience based on actual real life incidents, the reader can relate to them. People often misunderstand the link between real life and theories. When we feel somebody in impractical, we will say – 'he is too theoretical' of that 'he lacks understanding of reality.' What we forget is that a theory is nothing but generalisations drawn from real life experiences; the two, reality and theory, are organically linked using deductive logic. Mr K Sekar emphatically shows this in his book, whether it is from the sword hanging over his head when the barges he loaned to Airport Authority when building the Agatti airstrip drifted away, or for buying bags at 50 paisa more when building the slipway at Port Blair. He also uses various incidents from the Mahabharata and Ramayana, as well as tales from elsewhere like the one about the king of Amarapoora (Burma). All these to make a valid point that would help the reader better face professional and technical challenges and problems. What I particularly related to was the story about how the same medicine does not work for stomach ache of different persons, the objective conditions must be understood and blind copying or one-size-fits-all is a remedy for disaster. Also, that one must be aware of one's limitations and capacity.

I myself have gained a lot from my extensive interactions with Mr Sekar, which has got further strengthened from reading his manuscript. It is never too late to learn for all of us! I commend this book, particularly to the youth, in whose hands lie India's future. Even as they acquire more, and specialised technical, knowledge, they should learn from experience, their own and those of others. This book is a good place to start from.

Shakti Sinha

Note:

Mr. Shakti Sinha was a retired senior officer from Indian Administrative Service (1979 to 2013)

He was P.S to hon"ble Former Prime Minister of India Shri. Atal Behari Vajpayee

He also served as Chief Secretary of A & N Administration

He is the Chairperson of an upcoming think tank, South Asian Institute for Strategic Affairs (SAISA), and is also the head of Policy Research Group at the Bureau of Research in Industry and Economic Fundamentals (BRIEF)

1.0 Prologue

"Never think through your mouth. Between your mind and mouth, have different filters so that you speak only what is required."

-LSS

While attempting to link my hands-on experience of 32+ years in projects, I propose to brief on the terms used in current management techniques for the guidance of those young aspiring managers. I propose to identify Management at four different levels with reference to the level of experience, exposure, hands-on learning, expertise gained through applications etc. The four groups of management are:

1. Novice (Fresh)
2. Middle Level
3. Executive Level
4. Top Management

Novice is a person who is fresh to his field without any hands on experience in projects or management but desires to grow.

Middle Level Managers are those who have come into being in projects and started applying or understanding management techniques and have gained an experience of around 5 to 8 years in their field. These are the type of professionals who act as a bridge between Novice and the Executive Level Management. They are expected to train and guide the Novice and also responsible to deliver the output to meet the target set by Executive and the Top Level Management.

Executive Level Managers are those who are involved in decision making exercise on projects and policies of the organization in close consultation with Top Level Managers. They are the monitoring and controlling authority of managing any project. They form teams for projects in consultation with Middle Level Managers as well as Top Level Management keeping in mind the economics of the project and the organization at large. These professionals will typically (not necessarily) have about 12 to 15 years of project management experience.

Top Level Managers are the leaders at the top positions of an organization and decide on policy matters, financial matters and on major contracts, jobs etc. Their main job is to prioritize projects and allocate resources.

With an understanding of these definitions, I share my experiences and experiments I have come across during my long service in infrastructure projects in remotest islands. These life experiences remain so very green in my memory and when these experiences reach my close associates with whom I shared them, I am sure to see the thrill and happiness in their faces and mind by their rewinding of their past times with me! For many others, it might be a pleasant surprise as they were not privy to several of these episodes as they were serving in different parts of the islands when these incidents occurred.

For other readers, this could be an eye opener who can understand the problems, difficulties created by location, people, environment, climate and colleagues and bosses too in several occasions especially when one works in Government Department and throw light on what to expect and what not to expect from your boss and how one should be prepared to face any contingency.

SECTION 2.0

NOVICE

2.0 Novice

"Happiness lies within you only when you can think and talk what is pleasant to others. At least avoid pointing out their mistakes without hesitation. Your diplomacy matters very much under such circumstances."

-LSS

It is always going to be thrilling when you are going to attempt something which is new for you. How better can it be explained? Well! It is a feeling you feel when you feel that you are going to feel something which you have never felt before! That is what a novice is. I was a novice at that point of time with all excitements to involve myself in the project, to understand, to dedicate all my energy with will, to prove myself fit for projects and to share my experiences with my near and dear!

After completing my post-graduation Master of Technology(M.Tech.), in Indian Institute of Technology, Madras, I joined as Lecturer in Civil Engineering Department in Siddhaganga Institute of Technology, Tumkur in the State of Karnataka and I was well received in the profession by the institution and most of all by students. But I always had a feeling that I had only theoretical knowledge and I should have hands on experience on project planning and execution. By then, I was responding to an advertisement of Union Public Service Commission of India (U.P.S.C.) calling applications for the post of Assistant Engineers in Andaman Lakshadweep Harbour Works under Ministry of Shipping and Transport, Government of India. Soon, I was called to attend an Interview in U.P.S.C., New Delhi for which to and fro train fare was reimbursable. Basically, I was thrilled that I was going to travel free of my cost and I was to see Delhi, the capital of India for the first time. When I applied for leave with my department for proceeding to Delhi, my Professor discouraged me in shifting away from the present job stating that I had glued very well with the job and my students too liked me so much. But then, I convinced him that it was an opportunity for me to see Delhi free of travel charges and that I was only going for the interview and if at all I got selected, we can think about that later. I got permission, attended interview, did very well, went around Delhi without even knowing Hindi language (thanks to politicians of Tamil Nadu, the state where I had my education, who had vehemently opposed entry of Hindi in Tamil Nadu while their own family members who are in rich position today are capable of speaking Hindi and many other languages but they still oppose Hindi for their own benefit.) One fine morning I got appointment order from A.L.H.W., Port Blair asking for my willingness to join as Assistant Engineer for the Construction of a new Dry Dock Project at Port Blair. This was what I was yearning for- hands on project experience! Finally that was in my hands. After a great deal of persuasion and requesting, the college agreed to and relieved me with an assurance from me that in case I didn't like the place or the job, I should return back to them without fail for which I agreed. After that, there was no going back at all!

2.1 Starting of My Journey in Project Management

> *"Never ask for "Respect". It is not a thing to be demanded. It has to be earned by your decency and diplomacy in handling the situation."*

> *- LSS*

On 26th, December, 1978, I joined as Assistant Engineer in A.L.H.W., Port Blair and got posted at construction site. Before going into details of the project, let me first share some vital details of A and N Islands.

It is an archipelago lying in Bay of Bengal at a distance of about 1200 km from mainland Indian ports, Chennai and Kolkata. These islands are stretching from North to South as Emerald drops in the sea consisting of about 520 islands, islets and out crops. But only 34 of them are inhabited and the total population all put together was only about 150000 those days. (As per 2011 Census, the population is about 380581.) Out of the total land area of about 2600Sq.km, 92% of the area is under forest cover! The islands look like "Paradise on Earth" due to their green cover, unpolluted air and clean roads, which is due to the heavy rainfall which is about 3000 mm /year. Port Blair terrain is hilly with good roads and excellent view of sea from most of the locations and the sight of small water crafts always moving here and there with their sharp wooting noise and mellowed down dud dud dud noise of their engines from a distance ! Those days,there were only five passenger transport buses but there were good number of scooters moving around with people wearing very contrast color dresses like red, yellow, green etc. Fewer street lights! Small bazaar area! Main attraction in the bazaar was one Kattabomman Hotel which serves iddlies, dosas with sambar, chutney and a fat dark chap extending all courtesies once you enter !

There was no flight service to Andaman those days. All passengers and cargo were to be commuted invariably through ships which was available at an average of two sailings / month from Chennai or Kolkata to Port Blair and vice versa. In case of commuters to other islands, they have to wait at Port Blair till there is a sailing planned and announced through All India Radio, Port Blair in their news bulletin which almost 99 % population there was depending upon. There was no television, no telephone in most of the places, no 24 hrs power supply as the power in various islands are captive DG set generation for the local requirement which is dependent on supply of diesel. Whenever the inter island vessels touch outlying ports, say Little Andaman or Katchal or Kamorta or Camp Bell Bay in the Southern Group or Rangat or Mayabunder or Diglipur in the Northern Group of Islands, most of the population of the respective island would be in the harbor area which would have darned a festive atmosphere as that was the only entertainment for them. People offering and drinking tea, chewing pan and spitting in all available corners, small business men trading with ship crew for illegal bottles of whisky, rum etc. All essential commodities like vegetables, fruits, rice, cereal etc. were also transported by these ships only to inter islands. This back ground information would make it clear to any one as to why these islands were called "KAALA PAANI "which means "God forbidden Place ". Under such conditions, my journey of Engineering started at Port Blair in Dry Dock site!

For those who must be wondering what a dry dock is, I will give a short description of it. It is a concrete structure in the form of a rectangular box with its front side openable by fitting up suitable steel type gates, the bottom floor of the dry dock sufficiently lower than ground level to let sea water enter into the dock to facilitate entry of ship into the dock in floating condition. Once the ship is floated inside the dock, the gate in the front is closed by electrical or mechanical power against the side walls of the dock to provide water tight situation. Then the water inside the dock is pumped out by operating heavy duty dewatering pumps. The ship which is floating inside the dock slowly sits on the pre set keel blocks so that the keel portion (the bottom most edge portion of the ship) is accessible from below for carrying out the repairs with the dock almost dry. This is why it is called a dry dock. Once the repairs are completed inside the dry dock, the gate is opened to let in water and the ship starts floating and slowly pulled out of the dock. This is how a dry dock functions in general.

Dry Dock at Port Blair

This project involves various parallel activities like construction of cofferdam, casting of pre cast R.C.C. piles for the cofferdam,(to explain in simple terms, piles are concrete pillars with steel reinforcements which are fabricated and cast separately as readymade pillars which are lifted, placed in position with its sharp edge touching the ground at its location of driving and hammered down with heavy steel solid hammer of weight about 3 or 5 Tons), dock floor(which is the concrete floor forming the floor of the dock at the deepest portions of the dock pit which is constructed with steel reinforcement bars and rich concrete to take the load of the ship when docked inside for repairs) etc., procurement of materials, machinery, excavation of earth for the dock floor with huge earth moving equipment and cranes etc.(As the dock floor has to be much below the ground level to facilitate flooding of sea water and to allow the ship to float from sea to the dock, huge quantity of earth has to be excavated from the dock location to the required level to form the dock pit employing large earth moving machines).

The incident what I am narrating now happened during construction of cofferdam. Cofferdam is nothing but construction of a seemingly impervious bund at the entrance portion of the dry dock (a temporary dam like structure at the mouth of the dry dock project connecting land and sea) constructed on the sea side to facilitate construction of the project in a relatively dry situation. (As the project area is just adjacent to sea, it is practically impossible to stop seepage of water in the project area, especially when the earth work excavation goes at least 10 meters below the ground level and definitely below sea level.)

2.1.1 Thinking Different

> *"Never react to anything instantly. Absorb into your brain, understand what is said, analyze and respond that too only when what you say is palatable to others."*

> *- LSS*

An interview was under way when the person interviewing was in a hurry. He told the person sitting in front of him who came for attending interview,-

"Gentlemen! I have three questions for you. Please listen to them carefully, take your time and when I come back in ten minutes, you can answer them. Okay?

First question: How many days in a week start with the letter T?

Second Question: How many seconds are there in a year?

Third question: What is the first and second name of God?

I will be back in ten minutes. Think for answers and be ready. All the best!"

He left the hall with a funny smile on his face.

The person took out a magazine on the center table and started reading easing himself comfortably on the sofa.

Soon, the interviewer came back and was slightly shocked at the stature of the person inside with magazine in his hand and enjoying reading it too!

"Well Gentleman! Are you ready?"

"Yes Sir. Answer for the first question is TWO.'"

"Okay, Can you say what they are?"

"Yes Sir. They are TODAY and TOMORROW!"

Interviewer got a shock and said," Well! Your answer is different but sounds logical and so I give credit to you. Go ahead and answer the next question."

"Sir, the answer for the second question is TWELVE."

"Do you mean Twelve seconds in a year?"

"Yes Sir."

"Can you explain?"

"Yes Sir, January Second, February second, March Second….. up to December Second, which ultimately makes up to twelve seconds."

The interviewer was scratching his head and said," Though that was not what I expected, your answer seems to be logical. So, I give you marks for that too. And what is the answer for the third question?"

"Sir, the answer for the third question is: First name of God –"**Oh**", the middle name of God –"**My**". That is how He is **Oh My God**!"

The interviewer got astonished of these answers and offered him the job saying "You are thinking differently and I am sure you are going to be a great success! Congratulations and all the best!"

This is the quality required for anyone who is engaged in any project as thinking out of the box will lead to success. Without studying any management books, I was experimenting on it when I was a Novice in the project scenario of Dry dock project at Port Blair.

The activity of driving of pre cast R.C.C. pile (Reinforced Cement Concrete pile) for the cofferdam was in progress when the then Chief

Engineer and Administrator of the department Mr.G.Raghavan came on inspection tour to Port Blair. As dry dock project was the hot and important project for the department, he came to site along with other senior officers. It was then one pile driving was completed. As I was the new recruit, I was introduced to him. He started talking to me directly.

CEA: Mr. Sekar, Welcome to A.L.H.W. Study the project details, site conditions well and we want results from you. Well! Tell me when are you going to organize driving of next pile?

OLD TRADITIONAL TIMBER PILE DRIVING FRAME

Me: Sir, in about 45 minutes, we will be able to drive next pile.

CEA: What? Did you say 45 minutes?

Me: Yes Sir.

CEA: Are you mad? Do you know that the pile driving frame (which is as heavy as 8 tons) is to be moved to the next pile location which takes at least two days?

Me: Sir, Believe me. We will do it!

He gave me a peculiar look and told other officers in a serious tone that the new recruits are raw from college and unless they are trained well it may be difficult to complete the project in time. While he was busy with other officers, I spoke to my team to organize for next pile as we planned and to inform me at the site office as soon as they are ready. I joined CEA and others and moved with them to the site office there. We had detailed discussions on the project and reviewed the action plan with reference to targets and achievements. In about 40 minutes, mu junior engineer came to site office and informed me that the next pile is ready for driving. CEA looked at me with lots of surprise and said, "Come on! Let us go there and see what you people are doing."

When we went to the site, everything was ready for driving the next pile.

CEA: Tell me Mr. Sekar! What did you do?

Me: Sir, The earlier practice was to move the frame manually which took at least two days time. We at site have the Diesel Engine Double Drum Winch which is used for driving the piles. We just thought why we cannot use this available power winch to move the frame too to the required location so that we can save quite a lot of time. We tried with all safety measures and that's all Sir.

CEA: That is great! I am so happy the way you people thought and acted. Congrats. Keep it up! Never compromise on quality and safety!

Our team at site was very happy that our work got recognized and appreciated.

The person whom I am indebted to for inculcating this aspect of thinking differently is Mr.B.V.Balasubrahmanyam, who was my boss when I was serving as Assistant Engineer in Camp Bell Bay, Great Nicobar Island in the early eighties. Once we were encountering certain problem at construction site and all of us were discussing the solution for it. Out of curiosity and immaturity then, I jumped into the arena and explained about a similar problem which I had experienced in my earlier project and suggested the same solution which we had applied earlier. My boss with all humility and concern, explained to me the finer differences in the current situation and also driven the thought strongly into me that one should not jump to conclusions without properly analyzing the ground realities and all the more important message was that every problem should be approached with open mind and one should think out of the box for better solutions.

He was also citing an incident which was an eye opener for me. The incident was:

When the United States of America was to launch the first rocket with astronauts into space, there were rigorous exercises in the simulated modules. They suddenly encountered a serious problem. That was, when the astronauts were to log in their record of observations, they could not write anything on paper as the pens they were using were not writing. There was panic all of a sudden. They realized that as the simulation module which is a replica of space conditions was lacking in gravity and hence the ink in the pens was not flowing to the tip of the pens. There was a high level meeting which finally came to a decision that The Research and Development Wing of the Space Department would invent a new pen which would write in zero-gravity conditions and a few million Dollars fund was immediately allotted to them.

After the meeting, the Chairman of the Committee returned to his office with all tension in his face. His lady steno was told about the problem. Immediately, she told, "Sir, in such case, I can spare a packet of pencils given to me."

The Chairman looked at her with all surprise and appreciation!

This is just an example of how one should think out of the box.

This acquired virtue was useful to me later in my life. My daughter Jyothsna, when she was three years old was studying in Pre Kinder Garden in Toc H Public School at Kochi, Kerala. The education department in the state of Kerala was conducting various state level competitions for juniors of different categories. For pre KG level, there were many streams including elocution competition. The topic for elocution was "My Mother". It was such a difficult topic for the competition (as it is so simple for elders). My wife was particular that our daughter should participate in it and should also win. You can understand the pressure on me. Firstly, the topic is so straight that everyone were expected to say something like "My mother is the best, she feeds me what is good for me, she gives me milk, she takes good care of me etc." What else a child can say from its perspective? In any case, that would be the perception of any normal parents to teach their children I thought.

Secondly, a pre KG student did not know to read of his or her own. So the matter was to be fed into her mind somehow.

Considering all these aspects, I mulled over the situation and thought that I should do it differently. So, I prepared a script something like this:

Respected and Revered Judges, Dear loving audience and my dear young friends!

"It is a great privilege for me to stand before you, NOT to give a speech on the topic "My Mother" but to share my tender feelings when I think of my mother. She is the embodiment of love, affection, care, minding etc.

I have never been hungry and I do not know what hunger is. She only knows when I could be hungry and feeds me in time. When I didn't eat properly, it was not me who felt weak but it was she.

Whenever I fall down and injure myself, it is not me who feels the pain but it is she.

Whenever I fell ill, it was not me who was suffering but it was she.

Respected Judges and my dear loving audience! I really don't understand why it is like this. Perhaps, you may know it better to explain to me.

I love my mother very much than anything else in the world."

By enacting this repeatedly to my daughter, she learnt it by heart and presented so nicely on the stage.

Out of about eighty contestants, she was declared as the winner! Winning a state level competition, that too in Kerala where competition is always tough was a real height achieved, perhaps due to the application of "Think Differently" postulate which had a triumphant victory!

2.1.2 Split Second Decision

> *"Never ever expect people to "obey" you on whatever*
> *you say. It all depends on the veracity of your expression*
> *and your capacity to impress them."*

> *-LSS*

I would like to take you to another breath taking incident happened in the same project. We were having an idea of huge earth work involved in the project and that heavy equipment were used including cranes to excavate earth from considerable depth below ground level. The extent of excavation was to a length of about 120 meters by length, 30 meters by width and to a depth of about 10 meters below ground level. Of course, providing mild slopes to avoid slip circle failure of the standing earth. (Slip circle failure is nothing but sliding of the earth if proper slope was not maintained on the sides while excavating. The process of sliding of the earth into the pit due to poor holding capacity of the column of earth standing vertically). Periodically, the payment was to be made to the contract agency who was executing the job of earth excavation. Every time, measurements were to be taken by the site Engineers by using leveling instruments and level staff which is a graduated long staff. (One could have seen on the road side of new highway road constructions where a camera like instrument is mounted on a tripod with an engineering supervisor standing behind it and seeing through it towards the calibrated staff.) This staff was to be held by a person at different points of the grids by moving from one location to another and the Engineer records the measurements by seeing through the leveling instrument. At one point of time, such measurements were being taken at a depth of about 7 meters below ground level where the complete area was so slushy because of the proximity of the sea. When the person holding the staff moved to the next location, he started howling "Help. Help "in a very loud voice! Everyone around was shocked to see that he was slowly sinking beneath the slush, which was acting like a quick sand. I was standing just on top of the dock pit when I heard this cry. I understood what was happening. I looked around and saw a few timber pieces lying around used for shuttering. I picked up a few pieces and threw the same to the sinking man and shouted at him "Hey! You put the staff vertically on the timber piece and heave yourself up! "He did exactly the same and could arrest his further sinking into the mud. Soon others rushed there with ropes etc., threw it and he was pulled up to safety. His eyes thanked me so profusely that even words can never match. This incident gave me an immense satisfaction and happiness which I used to nourish in my memory.

2.1.3 Localized Innovation

> *"Do not expect that it is the responsibility of others to keep you happy. That is a disastrous attitude. You can never be happy."*

> *-LSS*

One more incident in the same project brought me appreciation which stands as an example of application of simple logic and positive approach to the problem with open mind by innovating with localized resources. The pile driving for the dock floor was to start by making use of the same pile driving timber frame which was used for the cofferdam works. While the pile for dock floor was lifted, this frame started twisting dangerously.

Then we all realized that the weight of this pile cast for driving in the dock floor area is more than that of the pile used for cofferdam as the length of the dock floor piles were more. All Senior Officers came to site and started analyzing the situation and to sort out the problem. Some of my colleagues suggested to go for a new frame, some suggested for a steel frame etc. When CEA asked my opinion, I suggested that we might consider adding additional diagonal bracings with adequate size timber beams at each stage of the existing timber frame which might take care of the torsion (twisting) effect which could be done in a short span of time with locally available materials, labour, without much of additional cost. This proposal was unanimously accepted and this proved to be extremely effective which had led to accelerate the speed of execution thereby avoiding Time and Cost Overrun.

Multiple pile driving in Andaman Islands

2.1.4 My Experiment on Sequential Approach:

That was the time when I was shaping up in the project and my Chief Engineer and Administrator was one Mr. Sivaswamy. When I joined the department as Assistant Engineer, Mr. Sivaswamy was an Executive Engineer to whom I reported directly. As he knew me well from day one of my service, he called me to his chamber one day and said "Sekar, we are formulating a plan for incorporating extraction of wave energy into the project of construction of a break water at Mus Harbor of Car Nicobar for which huge concrete caissons (huge concrete boxes which are pre cast, floated) are to be fabricated elsewhere and towed to the location and sunk in position. You may think seriously on this project, find out possibilities, various options and work out something and come for discussions. Can you do it?"

I replied in affirmative and returned to my work space. As several questions came to my mind, I started jotting down all my questions on a paper.

What is wave energy, how does a concrete caisson look like, where was it done earlier, what is the performance of the same if there is any, what are the components of a caisson, what was the capacity of power proposed to be obtained through this wave energy proposal, how many caissons are required for obtaining the target power, what is the requirement of construction space, is it available in the island of Car Nicobar, if not at Port Blair, If not at a location in mainland which is near a harbor and also at a place where construction materials are available, how to construct the yard for fabrication of caissons etc. I started working on it day and night.

Slowly things started falling into place. I collected details of caissons, derived the dimensions of each caisson, calculated the space requirements for the casting yard. Then I realized that there was no space at Car Nicobar or even in Port Blair. It was found that the Tuticorin Harbour on the Eastern Coast of India was best suited for the same as there were sufficient stone quarries around the area. I started working out every section of the project separately and prepared documentation chapter wise and approximate costing too as the supplement of each section. Then the major component was towing of each caisson to Car Nicobar from Tuticorin Harbour and sinking the same in position. That was beyond my capacity to assess the possible agencies that could do it and costing of the activity. So, that part of the work was left with blank. With all other details compiled almost in the form of a big project report, I went to him after 20 days and gave the book in manuscript form to him.

He looked at me once with disbelief and started turning the pages. After perusing through the abstract I presented to him, he asked me to take a seat in front of him. He got into reading more and more seriously with more attention. After a few more pages of scanning, he rang the bell and ordered for coffee for both of us. When coffee came, he offered it to me and told his Personal Assistant that no visitors are to be allowed unless he called again. With his entire mind into it he was going through page after page and there was pin drop silence there for about twenty minutes except the noise of turning of the pages. When he reached the section of costing of towing of caissons, he looked at me. I pre-empted, "Apologies Sir, I was unable to conceive this portion of the report and actually wanted to take your advice on this, which is why I am here now."

He turned to me ecstatically and said, "Sekar, I never thought that you would go so much into details on this project. You have done a great job. In fact I thought of getting some startup details from you which I can give to some premier institutions like Indian Institute of Technology Madras for working out these details what you have done now. I am proud of you."

I was so happy to hear that from a capable, strict and serious officer and responded, "Sir, I am also a product of Indian Institute of Technology Madras. I completed my post-graduation there."

When he presented this project report to Ministry, he called me and showed the page in which he had mentioned my name prominently as resource personnel.

Though he went on retirement 20+ years ago, we are still in touch even today and I always hail him as my friend, philosopher and guide. When we met recently, he was again remembering this episode and reminded me. I again expressed my gratefulness towards him with all humility.

A Novice, who wants to be successful in a project should have the inquisitiveness to know the project in detail and tries to learn the micro aspects at every available opportunity. He proves himself to be a perfect "**Learner**". His learning not only starts from his office but extends to going in search of related books, journals, papers etc. and tries to increase his "**Input**" capacity so that he will equip himself well to participate in useful and fruitful discussions wherein he can contribute. Once these aspects are imbibed in him, he starts "**Believing**" in himself and others which gives him "**Confidence**". The most important aspect that a novice should practice perfectly is "**Adaptability**" to local conditions, environment and the people connected to the project. He should be "**Positive** "in his thinking and approach applying his diligence and always try to display "**Discipline**". These essential qualities required for a novice are discussed in detail in the coming pages.

2.2 Essential Qualities of a Novice

> *"Anger is a lethal weapon for self destruction. It may relieve your pressure temporarily. But, when you reach your normalcy, your near and dear would have distanced much away from you."*
>
> *-LSS*

As seen in the earlier chapter, the essential qualities required for a novice to groom himself up as a good manager in the future are:

1. Eager learner
2. Ability to Analyze Inputs
3. Belief in oneself
4. Self confidence
5. Adaptability
6. Optimistic
7. Self discipline

These aspects are discussed in detail as below.

2.2.1 Eager Learner

> *"What you may say may be reasonable. But the tone of your saying makes all the difference. It is better to be a sweet poison than poisonous sweet."*

> *-LSS*

A teacher in a primary school was teaching her students the importance of following Rules of the Road through a small story. She explained as 'Dear children, last week, one boy by name John, 8 years old was riding his bicycle on the road at a high speed though his mother was constantly advising him not to speed in his bicycle. This boy never heeded. That fateful day, a passing truck dashed his bicycle and he was thrown out on the road. He was lying on the road bleeding. He was rushed to a nearby hospital for treatment. His mother came running there and with tears all over she asked him, "My dear son, how many times did I tell you not to speed in your bicycle? See, you have broken your leg now. Will you promise me now that henceforth you will never speed like that? "John promised his mother that he would never speed and would follow all traffic rules. My dear children! What do you understand from this story of John? Please come up and tell whatever you have understood! "Some boys got up and told different types of morals they could imagine. Finally one boy got up and asked "Miss,

What happened to his bicycle?" His thinking may be right from a remote perspective for those who may argue that there is no reason to think that the boy was wrong. But was that the lesson the teacher was trying to convey to her students?

Our learning should be streamlined towards our vision very clearly. A novice learner keeps his vision to learn extremely focused. He would like to seek such roles where he can exhibit his technical competence. He would enjoy his experience of learning with all humility.

He would avail all possible sources of trainings and enrichment organized by his organization through which he can improve his skills. The true knowledge is knowledge of causes and that the process of learning consists not in what is brought to the learner, but in what is drawn out of him. Energy, natural optimism, the ability to be a good listener and a quick learner are important personal attributes. Deeper rethinking focuses on the nature of knowledge and the learner's relationship to it.

Bad times have a scientific value. These are occasions a good learner would not miss. The learner always begins by finding fault, but the scholar sees the positive merit in everything.

In his article titled "7 Undeniable Reasons Why Some People Fail Where Others Succeed" posted on March 24, 2008 by Kevin Geary in "pickthebrain.com" beautifully describes Seven undeniable reasons why some people fail where others succeed as given below:

"Success happens not by chance, but because you were given a chance and took advantage of it." – Kevin Geary. In his words he says:

To most people, being born in a free country is the greatest gift. To others, it's a fleeting thought. For the latter, I feel sorry.

Before I go any further, I must admit that not everyone will find success. There will always be those who sit around waiting for success to find them. There will be those who are simply not willing to achieve it. And then there's the fact that success would not exist without failure. All of these things create what we know; a world where success and failure are experienced by different groups of people.

Everyone in a free country has the opportunity to succeed. So why doesn't everyone succeed? Because success and failure are choices made consciously and subconsciously and failure is chosen by many for various reasons.

1. They Define Success Wrong

"Striving for success without hard work is like trying to harvest where you haven't planted." – David Bly further adds:

Do you believe that success is won, innate, or earned? The answer someone gives can tell you a lot about them, and why they are where they are.

Success is won: if you believe that success is won, you experience animosity and envy toward those you view as lucky or more fortunate than you. You also believe that success is out of your control; it simply depends on a flip of the coin or certain circumstances.

How hard are you willing to work if you believe that success is won rather than earned?

Success is innate: people who believe success is innate often feel the same as those who believe it's won. The only difference is that believers in innate-success have a more pessimistic view of opportunity; it's trivial to them (we'll go over this a little later). Why does opportunity matter if success is innate?

How hard are you willing to work if you feel your opportunity doesn't matter and your chances of success are nil because of your circumstances?

Success is earned: the last group of people believes what we know to be true based on statistical analysis; success is earned. These people understand that in order to succeed, they must earn it. How do they earn it? They climb the mountain and utilize the same process others have used to achieve.

How hard are you willing to work if you believe success must be earned?

2. They Define Opportunity Wrong

"The ladder of success is best climbed by stepping on the rungs of opportunity." – Ayn Rand continues:

Do you believe that opportunity provides a possibility of success, a probability of success, or that it's trivial?

Let's ask the same questions we asked when we discussed success:

- How hard are you willing to work if you believe the opportunity you were born with is trivial?
- How hard are you willing to work if you believe the opportunity you were born with is a possibility?

- How hard are you willing to work if you believe the opportunity you were born with is a probability?
- I hope this is coming together for you. I still want to go further though. I want you to see exactly how your views on opportunity and success work together to help determine your outcome.
- People who believe success is won see their opportunity as a possibility, but sometimes as trivial.
- People who believe success is innate see their opportunity as trivial.
- People who believe success is earned see their opportunity as a probability, but sometimes only as a possibility.

In layman's terms, the rich see success as earned and view their opportunity as probability. The middle class see success as earned and view their opportunity as possibility, but sometimes as probability. The poor class sees success as won or innate and views their opportunity as trivial, or in some cases as a possibility, but not a probability.

Of course, people don't stay in one class their entire life. The people who move between classes tend to have the same outlook as those of the class they move to.

3. They Define Work Wrong

"The value of a man's position is often determined by the number of people qualified to fill it." – Kevin Geary explains:

We just discussed two important terms: success and opportunity. In order to continue our discussion further, we must discuss another, "work."

"But success doesn't always come from hard work!"

Inevitably, people will point out that factory employees work harder than CEOs. Of course, this depends on your analysis of the word "work."

Choose a corresponding term:

- Physical Labor
- Mental Labor
- Labor

Those who claim that success doesn't always come from hard work only acknowledge one aspect of work, physical labor.

Of course, work is labor, period. Excluding mental labor from the term work is biased and unfair. CEOs may sit at a desk, wear a suit, and enjoy the air conditioning, but that doesn't mean they labor any less than the man in the shop room, it's simply a different type of labor. Not accepting this is like making the argument that one who hates their job labors more than one who enjoys their job and the pay should be altered to make up for it. You see where this is going?

In terms of pay scale, people who run companies are worth a lot more than those who assemble products. Why? Because it's easy to find people who can assemble products and it's not very easy to find people who can operate multi-million dollar companies for a profit.

Needless to say, the man in the shop room wouldn't have a job if the CEO behind the desk wasn't doing his (and vice versa). The only difference is which job you'd rather be doing, and that depends solely on the choices you make throughout your life.

How do you think the CEO views success and opportunity? How do you think the shop worker views those same terms?

4.   *They Defeat Themselves*

"To expect defeat is nine-tenths of defeat itself." – Henry Mencken
says:

While there is a minority of people who actually choose to fail, the majority that fail simply make poor choices or have a poor outlook. Basically, for the majority, failure is a choice but not a decision.

I can't possibly list all of the bad choices people make which lead them to failure, but a few to get you headed in the right direction are:

- Abusing drugs or alcohol / addiction.
- Not getting an education.
- Having a poor work ethic.
- Having a child too young or out of wedlock.
- Immaturity / laziness.
- Borrowing too much money.

And the list goes on, and on, and on.....and on.

Of course, there are also those things which are out of someone's control.

If you're born into an inner-city family and attend a poor school system, you obviously start out behind others. If you're handicapped, your road to success may be longer and more difficult. But none of this bars you from success; I'll elaborate on this later when we discuss circumstances.

Lastly, as our quote up top reminds us, many people defeat themselves simply by expecting defeat in the first place. They don't expect success and it actually becomes a self-fulfilling prophecy. For more on this, you might like my highly popular article; Your Life Sucks Because You Expect it to Suck (and 10 Ways to Improve it Right Now).

5. *They Think Failure is Final*

"Success is the ability to go from failure to failure without losing your enthusiasm." – Winston Churchill adds:

"But, hard work doesn't always equal success. Some people work really hard, but fail. They tried and didn't succeed."

Failure is a key ingredient in success. Those who don't achieve success most likely quit after their failure. Quitting, of course, is a choice.

If you were to follow in the footsteps of a successful person, you would likely pass the remnants of multiple failures. If you followed in the footsteps of a failure, you would find their lifeless future at the feet of their first opponent.

So the question is, how hard and for how long are you willing to fight? There are no shortcuts, statistically. The vast majority of millionaires are self-made and far too many lottery winners are broke and worse off than they were before they won the lottery. Why? Because wealth is about behavior and money doesn't protect you from failure.

If you want to succeed where other people fail, you have to step right over failure and keep walking. The people who don't make it let failure defeat them. Failure becomes their end result because they refuse to walk any further.

Look at it this way; if you aren't dead yet, there's still hope.

6. *They're a Victim of Their Circumstances*

"The first step toward success is taken when you refuse to be a captive of the environment in which you first find yourself." – Mark Cain tells further:

One of the biggest rebuttals given by non-achievers is that they are held back by their circumstances.

I don't think circumstance is a fair argument though. Yes, you may be subject to circumstances that make it more difficult for you to succeed, but that doesn't change the fact that you start with the same opportunity as others; the opportunity provided to you by living in a free country.

It's also important to note that some people handle circumstances better than others. For instance, you can't say that a handicap is a circumstance that prevents you from achieving when others with the same handicap have achieved.

Everyone has issues, circumstances, road blocks, etc. It's all about how you deal with your circumstances and how hard you're willing to work to overcome them. But the basics don't change; you're still in a free country and nobody is preventing you from achieving except for yourself.

Circumstance is also unimportant because it doesn't determine finality. For example, a trust fund baby can lose his fortune with a series of bad decisions just as easily as a child from the ghetto can acquire a fortune with a series of good decisions.

Don't be quick to judge others based on their circumstances. Instead, judge them based on their ferocity in overcoming those circumstances.

7. *They Take No For an Answer*

"Opposition is a natural part of life. Just as we develop our physical muscles through overcoming opposition – such as lifting weights – we develop our character muscles by overcoming challenges and adversity." – Stephen R. Covey explains:

First, you are given an opportunity. Then, based on that opportunity, you hatch a dream. And when you try to execute that dream, you meet your opposition. It is here on the battleground, facing the opposition, that success is either realized or lost.

Everyone faces opposition on their way to the top. The crack babies and the trust fund babies both have their own sets of problems. And you can't assume that one faces more opposition than the other; everyone's life and path to success is unique.

The one thing they do have in common is the opportunity for success. But, as you try to succeed, there will be people and circumstances around every corner that try to tell you no. The disability you were born with tells you no, your abusive parents tell you no, your pessimistic friends tell you no, your lack of self esteem causes you to say no to yourself, addiction tells you no, and so on.

The people who succeed are those who don't take no for an answer. They shrug off the pessimism, they choose better friends, they put up boundaries with their family, and they surround themselves with positive people and things.

To conclude, Success is possible for anyone who is willing to achieve it. There are many who want success, but there is a huge difference between wanting-to and willing-to. You have to be willing...

The other thing to remember is that your outlook and the way you define success, opportunity, and work play a large role in determining your outcome.

If you aren't achieving, the first person you should always look to first is yourself.

You have to be conscious of time, i.e., what to do and when. *The Holy Bible says, "To everything there is a season, a time for every purpose under heaven" (Ecclesiastes: 3:1).*It is time to understand the nuances of this for a novice.

Focus is not a new management theory invented now. In the great Indian epic Mahabharata, this is well dealt in the episode where pandavas and kauravas were learning archery from their kulguru Dronacharya.

Dronacharya,as a part of testing the talents of his students, pandavas and Kauravas, took them all deep inside a forest. He planted a wooden bird on top of a tree at a distance of about a hundred yards away. He addressed his students," My dear students! Do you see the bird on top of the tree over there ?"

They all said, "Yes".

"Now, Children! You have to shoot the bird by shooting an arrow only once!"

The students were shocked to hear that. They murmured, "At such a distance?"

"That too with single shot?" exclaimed one of them.

Acharya told them, that he will call one by one and ask a question. Depending on their answer, they will be given a chance or refused a chance.

This made most of them more nervous as our children's syndrome before going into examination hall.

He called Yudhishtra, the eldest of all first. He asked Yudhishtra, "My dear boy, what did you see there?"

Yudhishtra with a confused look at his Guru, said, "Sir, I see the tree, branch of the tree where the bird is sitting and of course the bird surrounded by leaves!"

Dronacharya told him, "You cannot hit the bird!" and he called Duriyodhana to come next and aim at the bird. Then he asked him," What did you see?"

Duriyodhana replied, "Sir, I can see clearly the tree, its branches, the bird and you too!"

Dronacharya told him to put down his bow and return to his place as he too cannot shoot the bird.

He called everyone one after another and every one almost gave the same reply in different words. And every one of them was sent back to his place. Finally, he called Arjun and told him to aim at the bird. Then he asked the same question. Arjun replied, "Sir, I can see only the head of the bird."

Dronacharya asked him again, "The wings of the bird?"

Arjun said, "No Sir"

Guru said, "Now you shoot your arrow!"

Arjun did it and the bird came rolling down the tree the next moment.

Guru said to all of them, "My dear children! This is what you have to learn in your life. If you fox a target, all your focus shall be on your target only. If you lose sight of your target, you may miss your goal".

This is a great management lesson one can draw from the great epic of India Mahabharata.

This is about making the new knowledge fit with what the learner already knows, not making it mean whatever the learner wants. Good learners change their knowledge structures in order to accommodate what they are learning. They use the new knowledge to tear down what's poorly constructed, to finish what's only partially built, and to create new additions.

In the process, they build a bigger and better knowledge structure. It's not enough to just take in new knowledge. It has to make sense, to connect in meaningful ways with what the learner already knows.

There's always more to know. Good learners are never satisfied with how much they know about anything. They are pulled around by questions— the ones they still can't answer, or can only answer part way, or the ones without very good answers. Those questions follow them around like day follows night with the answer bringing daylight but the next question revealing the darkness.

Knowledge is inert. Unless it's passed on, knowledge is lost. Good learners are teachers committed to sharing with others what they've learned. They write about it, and talk about it. Good learners can explain what they know in ways that make sense to others. They aren't trapped by specialized language. They can translate, paraphrase, and find examples that make what they know meaningful to other learners. They are connected to the knowledge passed on to them and committed to leaving what they've learned with others.

2.2.2 Ability to Analyze Inputs

> *"Do not decide what others have to do. You don't have*
> *any right for it. If you try that, you are a rejected lot."*

> *-LSS*

This is an extension of the characteristics of a learner. The thirst for learning doesn't stop for an inquisitive novice within his work environment alone. He tries to explore more and more details from books, journals, magazines, websites etc. and collects all relevant and valuable information in trying to apply in his project wherever applicable in the interest of cost and time saving. The search for new knowledge is going makes him excited. He finds it all the more interesting to share the same with his colleagues and his team. While doing so, you may come up with so much of information which may not even be able to apply in the immediate term but nevertheless, interesting. You will find yourself absorbing it in the hope that this will differentiate your knowledge and capabilities one day in the future. And hence the reinforcement of the old saying- Knowledge is Power! The knowledge thus accumulated serves as the novice's strength.

Once you have sufficient inputs, you will be on the lookout for a better position or better job where you crave to put in use many of the inputs you have acquired.

You will be all the more inclined to systematically sort out and make a data base of your inputs in such an order that you will be able to access it at any point of time when they are required. While acquiring inputs, you may come across various interesting sections of project intricacies and perhaps you may even find one such field more interesting and appealing to you and you may decide to pursue your search in that field and perhaps that could be you field of your specialization in future.

Such assimilation of knowledge and inputs will drive you to learn and obtain in-depth knowledge in the field and one day, even without your knowledge, you are going to be recognized as a specialist in the field.

This could happen only if you are looking for every opportunity to utilize your assimilated knowledge at the appropriate time by identifying right situations. You should position yourself in such a way that your peers and bosses recognize that you have valuable knowledge learnt from authentic proven sources. That will elevate your status to that of a specialist.

Finally, never hesitate to put forward your views backed by your research whenever you get an opportunity. Your performance can be appreciated and rewarded only if the comprehensive and abundantly reliable knowledge you have gathered can create a lasting impact for your organization or your clients.

Management utilizes all the inputs from physical & human resources productively. This leads to efficacy in management. Management provides maximum utilization of scarce resources by selecting its best possible alternate use in industry from out of various uses. It makes use of experts, professional and these services leads to use of their skills, knowledge, and proper utilization and avoids wastage. If employees and machines are producing its maximum there is no under employment of any resources.

2.2.3 Believe in yourself

"You have all the freedom in the world to decide what you want and do what makes you happy provided it is not dependent on anybody's contribution or participation compulsorily."

Believing in yourself is the key to success in life.

One who believes in himself strongly is valued as "Trust worthy person" by others. Your belief in yourself reflects your core values, strong will and "sure to perform" attitude. This guides you in the right direction towards consistent set of priorities.

In a similar way, you also recognize others who are also have strong belief in themselves. There cannot be a clash of ego that you are right and the others are wrong. Be sure that you accept the values of others even though you may differ from you. This acceptance level on others will take you to greater heights and better recognition. Remember you disagree with the point being made, not the person who made the point. You can always express your beliefs without being judgmental.

Your belief should be able to convince others on the intent, meaning and purpose of your work and the directions you would like to take by taking them along. While doing so, you will recognize how important their contributions are going to be and also make it crystal clear how important they are to you and your unconditional belief on them.

Make them feel and believe that what you are planning to do together as a team is going to make a significant difference to the society which they will remember forever. Make a special collage of those things, people who have inspired you to reach this height. Sometimes, when you are in your low spirits, you may look upon your collage which will boost your spirits and morality and makes you believe in yourself more strongly.

On your war to this height, you would have come across many intellectual persons. Make as many of them your close confidents, guides and friends from whom you can draw inspiration from and boost your morale.

Recognize the skills that you have and the good things about yourself. There are lots! You may not always recognize them but they are there. One way is to look for the things that you don't struggle with or to look for the things that people compliment you on (even if you aren't very good about accepting the compliment). When you look at the things you do well, you'll feel more comfortable doing other things too. This is the meaning of believing in you.

Set goals for yourself and meet those goals. Set out and do things. Just thinking about what you want to do will only make you feel worse about yourself for not even starting them. Once you do make a goal, work hard to achieve that goal.

Long ago, there was a battle about to take place.

One of the generals was talking about tactics with his team of officers.

An officer interrupted him and explained that he thought that the strategy was a waste of time.

"The gods have already decided who will win." he proclaimed.

"Are you suggesting that fate has decided the result in advance?" the general asked.

"Yes, I am." the officer responded.

The general took a coin out of his pocket and said, "So if I toss this coin and it comes up heads, we win, but if it's tails we lose. Is that how fate works?"

"Pretty much." said the officer.

The general tossed the coin and it came up heads.

"See, the gods have decided. We can't lose now!"

They went to their troops with the good news and the soldiers marched into battle with renewed enthusiasm.

After a glorious victory, the officers met in the general's tent to celebrate.

"Do you believe in fate now?" the general was asked.

The leader smiled, reached into his pocket and pulled out the coin to show to the others.

It was heads on both sides.

"No, I don't believe in fate, just the value of self belief. When the soldiers thought that we couldn't lose, I knew that we couldn't lose."

Sometimes, we think that the script has been written and that success is for a chosen few.

I share this adapted story with you today to encourage you to believe in yourself, to have confidence in your abilities and to launch yourself into the fray with energy and enthusiasm.

If you do, great victories will be won.

Instead of viewing your failures as failures, view them as learning opportunities. You did something wrong so that you know for sure it's wrong and now you'll have an easier time knowing what's right and so embrace it. As Edison said reminiscing about his invention of the light bulb, "I have not failed 700 times. I have not failed once. I have succeeded in proving that those 700 ways will not work. When I have eliminated the ways that will not work, I will find the way that will work."

Everything that goes wrong in your life is an opportunity to learn. Sometimes we feel like we shouldn't do something new because we might do it wrong. This is a vicious mindset. Instead, give yourself permission to try something, even if you may get it wrong. If you never try new things, you'll never be able to make progress.

If you're really having trouble seeing all the wonderful things about yourself, you can always talk to someone who loves you. Tell them that you're having a hard time and they will find a way to bring you a better outlook. Sometimes we have difficulty seeing the best things about ourselves, but the people that really love us will never struggle to see those things.

If a situation or a task is just too overwhelming, feel free to take a quick break to catch your breath and remind yourself that you can do this. Even if the break is just inside your head, it's okay to stop and acknowledge that you need a minute to build back your determination.

Recognize that just as you make mistakes, everyone else makes mistakes too. And just like everyone else does remarkably well sometimes, you will do too. Everyone is equal and we are all worthwhile. We all have something to offer. Stop seeing yourself as lower or separate from other people and you'll automatically have a much better opinion of yourself.

Don't try to be someone you're not or be upset with yourself for the things that you think are wrong with you. There is a saying – "Be yourself, everyone else is taken!" You can try to constantly improve certain things about yourself, in ways that really matter to you. But you won't be happy if you force yourself to be someone or something that you are not. Try to love yourself for what you ARE and you'll find yourself living a much easier life.

Here is a true story which narrates how your strong belief in yourselves would bring you the reward of success. A man by name M. Farouk Radwan, turned the impossible to possible by his sheer belief in himself.

To start with his ambition, he was telling everyone that he wanted to become a millionaire and was discussing his plans with whomever he met, which was a big mistake.

With high aspirations, he joined a multinational company and started working in it and he was proud to be in the company. A few months went by. The initial charm started to fade away slowly and he started wondering what he was up to. He had to follow timings of the company every day, their rules and restrictions strictly and suddenly he realized that it was not the way he was planning his future dream. He was missing his home, his friends, his food etc. One fine day, he decided to quit the job and return to his village to start working on his ambition. People around him started laughing at him that his ambition to become a millionaire and his decision to go back to his village were quite poles apart. But he was stubborn on his decision, came back to his village and started working seriously all by himself without discussing his plans with anyone any more. He learnt the online trade thoroughly, developed his own web site for it and launched his business successfully.

Within a short period of time, he has developed his online business which allowed him to sell his products everywhere in the world sitting in his own place, working at his convenience and yet not compromising his meetings with friends. He did achieve his goal of becoming a millionaire!

Hence, let us all realize that nothing is Impossible in this world! Once you believe in yourself strongly, then no force on earth will be able to stop you no matter what restrictions you find along your way.

Get to know yourself before you believe in yourself. Know what works best for you. Also get to know your personality that is the most important thing to do if you want to believe in yourself.

2.2.4 Self Confidence

> **"The best way to command respect is to be courteous to others and respect them irrespective of their level or position. After all we are human beings."**
>
> **-LSS**

Here is an interesting story to demonstrate self-confidence acquired by a boy through an old man who is a motivator.

The boy approached a wise old man who is good at giving counsel. The boy explained the old man that he was feeling desperate and aloof amidst several people during meetings or discussions as no one was minding him and no one cares to listen to him and so he was so much depressed. He was seeking the advice of the old man.

The old man smiled at him and called him to come close to him. He murmured something into his ears. The young boy's face got lit with energy and vigor. He thanked the old man and left.

They both met again after six months as agreed earlier.

The boy hugged the old man and thanked profusely for his advice. He said, "Old man! It worked finally and has given thumping results! Now, everyone is so keen to watch my words and I feel extremely confident in the crowd as everyone pays attention to me. Thank you very much!"

The old man told him to continue practicing what he told him earlier.

"But", said the boy, "I am already grown very confident. Why should I continue that exercise?"

The old man said, "My dear boy! You should practice it further to ensure that you retain your confidence even when people fail to pay attention to you. That would be the real self confidence."

Self-confidence begins with 'Self'. It has far less to do with what others think of us than with what we think of ourselves.

Although we can never derive a solid sense of self-worth from our appearances, looking our best can provide a quick booster shot of confidence when we're not feeling so hot. For example, if we're nervous about giving a presentation at work or school, we can dress in a favorite outfit that we know looks great on us.

Being confident doesn't mean having unrealistic or superhuman expectations of ourselves. On the contrary, we are most likely to feel confident when we know both our capabilities and our limits, and can function within these boundaries. If we truly want to become confident, we must first recognize that confidence isn't about being or feeling larger than life. Confidence means that we are right-sized: feeling good about the abilities we have, neither downplaying them nor blowing them out of proportion.

Becoming confident is part of the lifelong process of discovering who we are as individuals. When we tune into our deepest dreams and passions and take steps toward making those things happen, we are then most aware of our purposes for living. Increasing confidence is a natural byproduct of following our hearts in this way – the feeling of confidence functions like a compass needle pointing north, signaling us that we are indeed moving in the right direction.

The most important step in building self-confidence is simply to take action. Working on something and getting it done. Sitting at home and thinking about it will just make you feel worse.

You gain strength, courage and confidence by every experience in which you really stop to look fear in the face. You are able to say to yourself, 'I have lived through this horror. I can take the next thing that comes along.' You must attempt things you think you cannot do.

When you do such things, you don't just build confidence in your ability to handle different situations. You also experience progressive desensitization. What that means is that situations – like for example public speaking or maybe just showing your latest blog post to an audience out there – that made you feel all shaky become more and more normal in your life. It is no longer something you psyche yourself up to do. It just becomes normal. Like tying your shoes, hanging out with your friends or taking a shower.

One important key to success is self-confidence. An important key to self-confidence is preparation.

According to research at Najafi Global Mindset Institute, self-assurance is a combination of self-confidence, energy level and resilience. Managers must believe that they can successfully take on more complex work. They need the energy to juggle multiple roles, issues and tasks at the same time. They also need a healthy dose of resilience – the ability to bounce back when things don't go well. All three of these can amount to a big psychological leap at each turn in the leadership pipeline. That leap may feel too intimidating for a manager to accept. This is where we see talented employees staying in the same position for years, not because they really want to but because it feels safe.

Consider the example of a marketing manager who has been successfully leading 10 individual contributors for the last five years. He is skilled at working with his employees and managing projects. Now he is promoted to marketing director, where he leads four managers who lead distinct marketing functions and have a total of 40 direct reports. Even though he may have the right skills and abilities to do the job well, the drastically increased scope and responsibilities are intimidating. Until he builds his confidence in the role, his insecurities and concerns may affect every decision he makes through indecisiveness or anxiety. He may become demotivated by the steady stream of challenges, and may even come to regret accepting the promotion. Some level of doubt is normal; it is, in fact, healthy to avoid becoming a narcissistic leader. But "planful" development that creates ongoing coaching and mentoring opportunities for this manager can make the transition smoother and less stressful.

Confidence comes not from always being right but from not fearing to be wrong.

Building self-confidence is readily achievable, just as long as you have the focus and determination to carry things through. And what's even better is that the things you'll do to build self-confidence will also build success – after all, your confidence will come from real, solid achievement. No-one can take this away from you!

By doing the right things, and starting with small, easy wins, you'll put yourself on the path to success – and start building the self-confidence that comes with this.

Looking at your goals, identify the skills you'll need to achieve them. And then look at how you can acquire these skills confidently and well. Don't just accept a sketchy, just-good-enough solution – look for a solution, a program or a course that fully equips you to achieve what you want to achieve and, ideally, gives you a certificate or qualification you can be proud of.

Even the small ones. Don't wait to get perfect. Any progress at all is worthy of celebration. When you celebrate, you are telling the universe "I am loving this. Please give me more!"

The important part of self confidence is about being able to relax with uncertainty. To be 'cool' in a situation really means relaxing with not knowing how things will pan out. If you truly tolerate uncertainty, you can do pretty much anything.

Poor self confidence is what usually makes it hard for us to believe in ourselves. If you want to believe in yourself and the things you do, you'll need to build your self confidence. The most basic way to build your self confidence is by changing how you think and talk about yourself. Stop beating yourself up. Every time you catch yourself thinking or saying something bad about yourself, stop. Say a few good things about yourself and then think about healthy ways to work towards fixing things that you want to improve about yourself.

It's hard to be confident in yourself if you don't think you'll do well at something. Beat that feeling by preparing yourself as much as possible. Think about taking an exam: if you haven't studied, you won't have much confidence in your abilities to do well on the exam. But if you studied your best, you're prepared, and you'll be much more confident. Now think of life as your exam, and prepare yourself. So much so in projects too.

Such a simple thing, but it can have a big difference in how others perceive you. A person in authority, with authority, speaks slowly. It shows confidence. A person who feels that he isn't worth listening to will speak quickly, because he doesn't want to keep others waiting on something not worthy of listening to. Even if you don't feel the confidence of someone who speaks slowly, try doing it a few times. It will make you feel more confident. Of course, don't take it to an extreme, but just don't sound rushed either.

Empowering yourself, in general, is one of the best strategies for building self-confidence. You can do that in many ways, but one of the surest ways to empower yourself is through knowledge. This is along the same vein as building competence and getting prepared ... by becoming more knowledgeable, you'll be more confident ... and you become more knowledgeable by doing research and studying. The Internet is a great tool, of course, but so are the people around you, people who have done what you want, books, magazines, and educational institutions.

Self Confidence is the first step towards self assurance. I am sure you all will appreciate the difference between Theory and Practice, Planning and execution. I feel that Self Confidence is theory and Self Assurance is Practical. You all must have heard of the ex world heavy weight champion Mohammed Ali. Many of you would have read his famous auto biography titled "I am The Greatest". Personally, it was a great book I cherished which I very often refer to my staff during my lectures. Before any major boxing competition, Mohammed Ali in every press meeting, used to announce openly that he would knock down his opponent in Round No...... This announcement showed his Self Confidence. Once he had announced this, his responsibility to his commitment started falling on him heavily which drove him to work harder and harder towards achieving his commitment. And every time he did that, he won titles. This application of theory to practice is Self Assurance.

Confident people inspire confidence in others: their audience, their peers, their bosses, their customers, and their friends. And gaining the confidence of others is one of the key ways in which a self-confident person finds success.

The good news is that self-confidence really can be learned and built on. And, whether you're working on your own self-confidence or building the confidence of people around you, it's well-worth the effort! The inspiring messages which are of interest and relevant to the subject as I could appreciate are:

"Because one believes in oneself, one doesn't try to convince others. Because one is content with oneself, one doesn't need others' approval. Because one accepts oneself, the whole world accepts him or her." As per Lao Tzu and

"One of my rules is never to look sideways at what other people are doing but instead, do what I feel is right." As said by Annie Bryant in Worst Enemies/ Best Friends

A self-assuring prophecy is a prediction that directly or indirectly causes it to become true, by the very terms of the prophecy itself, due to positive feedback between belief and behavior.

Another way to build your *self-assurance* is by proving to yourself that you can take on any challenge and giving yourself things to be proud of. Good ways to do this include traveling or volunteering.

Once you're aware of your strengths, you need to identify your weaknesses—and with good reason. The mere fact that we understand who we are, for better or for worse, has been found to improve self-esteem. Though it's in our nature to self-criticize, we also sometimes lie to ourselves about our talents, the real price of certain less-than-ideal aspects of our lives, and our true character traits. Facing your flaws means looking at patterns in your life. Are you often accused of, say, stubbornness, indecision, or hot-temperedness? Do you end up in the same old conflicts with partners, friends, or coworkers? Figure out how you're contributing to those situations, and you'll likely identify what traits you need to address.

Your confidence can be affected by the way that people you respect salute who you are and what you do. (There's a reason we call it "a vote of confidence" when others give us a pat on the back.) Being able to receive constructive, positive feedback can help counteract negative thoughts and build confidence. Besides, if you respect someone enough to take their criticism to heart, it's only fair to also accept their praise. If you can't stop yourself from brushing off compliments, practice saying this in front of your bathroom mirror: "Thanks. I appreciate your saying that. I worked really hard, and the fact that you noticed means a lot."

Whatever your objective is, you must define it, create a plan for achieving it, and set up a timeline for its completion in small, measurable steps. Charting these achievements can boost self-esteem as much as attaining your ultimate goal, because, to put it simply, a success is a success, and the more of them you have, the more favorably you're likely to view yourself. When you take deliberate measures toward a goal, you'll have the fortitude to think—and do—big.

1. Define your destination.
2. Draw a map from here to there.
3. Set up a timeline for arrival.

Taking reasonable risks sets us up for success not just once but repeatedly. No matter the outcome, the attempt itself represents an accomplishment that can be built upon. Next time you'll aim even higher, think even bigger—and before long making bold moves will start to feel like second nature.

Having a passionate group of supporters fosters a sense of belonging, which can translate to a feeling of security—something that helps us build confidence. But there is often strength in numbers, and identifying the right people to root for you can make a significant difference in the way you approach challenges. Athletes competing in high-pressure situations gain a measurable confidence boost from being cheered on by their peers; it can literally help propel them to victory, according to researchers. It's important for you to feel just as encouraged as you go through life.

When designing management development processes, it is helpful to be mindful of human resources as just that – resources. Each person has a unique perspective and experience that can be helpful to a colleague. Our organizations are full of coaches and mentors who just need to be identified and invited to serve in this capacity. Connecting employees in meaningful ways with one another strengthens engagement across the organization. It also provides them with coaching and mentoring that can help them break through the self-doubts or fears that may be holding them back.

The virtue of self-confidence is a thing to show, a thing to display and a thing to be contagious which catches up with everybody around. So be generous to spread it.

Your capacity to display calm and cool at the time of disarray, confusion and panic pronounces you the best person of self-confidence. A person with self confidence always thinks on "what next" even at the time of worst calamity instead of lamenting on what had happened.

One businessman was in debt and couldn't figure out a way out of it. Creditors were pressing him. Suppliers were demanding for a payment. He was sitting on a bench in the park with his head down, thinking what could save the company from bankruptcy. Suddenly, an old man appeared in front of him.

-"I see that something is bothering you", - he said.

After listening to the businessman, the old man said: "I think I can help you."

He asked the businessman, what his name was, wrote him a check and was shoving it into his hands saying:" Take this money. We will meet here exactly in one year, and you will be able to return it to me at that time."

After that, he turned around and disappeared just as suddenly as he appeared.

The businessman saw a check in his hands for an amount of 500 thousand dollars, signed by John Rockefeller, one of the richest people in the world at that time!

'I could end all of my problems in no time!' – He thought. But instead, the businessman decided to put the check into his safe. Only one thought about his existence gave him strength to find a solution to save his business.

With the return of his optimism, he concluded profitable deals. He was able to commit a few big deals. During a couple of months he got out of debt and started to earn money again.

Exactly one year later, he returned to the park with the same check. At the agreed time the old man appeared again. And at the moment, when the businessman wanted to return the check and to share his story of success, a nurse ran up and grabbed the old man.

- I'm so glad, I caught him! - She exclaimed. – I hope he wasn't bothering you. He always runs away from the house and tells he is John Rockefeller.

Surprised, the businessman was standing there bemused. During the whole year he was spinning and building a business, buying and selling, convinced that he had half a million dollars. Suddenly he understood that it's not the money, real or imaginary, that turned his life up. It was his new confidence that gave him strength, to achieve everything that he had now.

Never give up on your goal because you never know how close you really are to success!

2.2.5 Adaptability:

"Never clamor on some body's mistakes repeatedly.
That will keep you at arm's length from others."

-LSS

Enjoying success requires the ability to adapt. Only by being open to change will you have a true opportunity to get the most from your talent. Among the skills that employers are looking for in a potential employee, adaptability is ranked the highest along with communication, interpersonal skills and a strong work ethic. Every company looks for a candidate who fits within the existing work environment and is able to anticipate, respond to and manage change on a day-to-day basis.

"Those who cannot change their minds cannot change anything." says George Bernard Shaw.

"The measure of intelligence is the ability to change." says Albert Einstein.

"It is a law of nature we overlook, that intellectual versatility is the compensation for change, danger, and trouble. An animal perfectly in harmony with its environment is a perfect mechanism. Nature never appeals to intelligence until habit and instinct are useless. There is no intelligence where there is no change and no need of change. Only those animals partake of intelligence that have a huge variety of needs and dangers." says H.G. Wells in his "The Time Machine".

You are sure to experience change in the workplace, regardless of your occupation. Modifications to management, technology, procedures, expectations, policies, and even company culture take place all the time. However, humans have a common, natural instinct to resist change. Unwillingness to be flexible and adapt may prohibit or even derail career advancement. You can become a valuable asset to any company by showing a willingness to embrace change.

Whether you are looking for a new job or are currently employed, staying open-minded is essential to career success. Particularly if you are searching for a job in this austere economic climate, demonstrating flexibility will help you stay afloat.

Being flexible is also about being versatile within your vocation. You must showcase your skills in a way that demonstrates your ability to accommodate the demands of work, as well as the needs of your co-workers and superiors. With seemingly limited resources in today's economy, employees must adapt to accomplish more with less, and work quickly through it all. In turn, this ensures employees benefit from being willing to adapt their work styles in new ways and with positive attitudes.

Moreover, you must learn to read your co-workers and supervisors so you can adapt your communication style in order to suit individual preferences. Some people dislike talking on the phone while others prefer face-to-face communication. Taking note and adapting to these differences will make your co-workers feel like you value them, thus making you a well-liked and winsome team member.

As an ardent and futuristic manager, you do have your own plans of your future- regarding your family, your profession, your line of interest in various fields and you have plans drawn as to how you wanted to proceed on each of them. But, there are so many external factors which are sure to influence at the time of implementation. A good novice, anticipate such circumstances and adjust himself so readily to take a detour of his ways and means but will always remains careful to ensure that he redrafts his plan in such a way that he could proceed well as the situation warrants without compromising on his ultimate goal. This in other words is the action of camouflaging himself into the new situation perfectly but still going great towards his goal. This process of pulling away from the initial plans and responding willingly to the situation is called Adaptability.

This adaptability is expected to be smooth and at no stage, the resentment of shifting from original plan should be visible. That should happen in a calm and reassuring atmosphere.

Adaptability is not a virtue that everyone can easily adopt, as the mental or ego system of most humans will not so easily accept it. So you may be a catalyst with your cool and level headed poise to help others overcome such a syndrome and guide them to real adaptability.

By doing so, you may also run the risk of some irritants abusing your cool temper as purely submissive for all changes imposed. So, you have to strike the balance between your temperament and being transparent about your boundaries to send the message that that you are not to be taken lightly and that you would accept no nonsense by any means. By doing so, you will gain emotional capital among your team members and also among others who have already started to look up to you for your capabilities.

Your adaptability helps you to share your views with likeminded professionals who will also be in a position to guide you for your future goal.

Most importantly, there may be occasions when adaptability renders adverse results. That is not the time to blame each other and to make a big cry. The theory of "crying over spilled milk" could be the worst thing to be accepted and to be avoided completely.

Adaptability in itself explains to change of plan in changing environment. Hence, there is nothing wrong in thinking of alternatives at every stage and when one doesn't seem to go well; you can propose alternatives as after all it is the question of adaptability.

Adaptable people are able to bend when their first suggestion or preferred solution does not go over well. Being adaptable involves preparing backup, alternative options for discussion. Within a work team, presenting multiple ideas and showing acceptance when your primary choice is rejected helps you come across as a team player. In making a sales pitch to a client, conveying alternative product solutions when the first recommendation is rejected can help you get more sales and achieve better results.

People who aren't adaptable tend to get stressed and uncomfortable when faced with new, unexpected and urgent projects. To show that you're a team player and adaptable, be willing to take on tasks or projects when urgency is key and the work is important to the organization. While you don't want colleagues to take advantage of your willingness, agreeing to these surprises from others gives you a better chance of finding them agreeable when you need immediate help.

- Create an atmosphere that is open, supportive and welcoming of new ideas.
- Help each individual to feel that they are valued members of a community.
- Encourage risk taking together with risk management.
- Some people (for example, those with autism) rely heavily on routines as a source of security and comfort and dislike sudden change. Let people know well in advance about changes to routines in order to let them adjust to the idea and to avoid upsetting them.
- For those who have difficulty focusing on appropriate detail or who are easily distracted, provide a working environment that is not too 'busy', e.g. by screening off desk space.
- Some people with Mental Health Difficulties may be affected by prescribed medication which can affect concentration or make them feel excessively tired; they may need more time and support to adjust to changing situations.
- People with autism frequently demonstrate misunderstanding or naivety within social interactions and may have difficulty interacting with their peers and teachers/trainers. Provide these individuals with clear routines and written guidance on procedures.
- One in four people in the UK experience Mental Health Difficulties at some point in their lives – do not write someone off because they happen to be ill at this time.
- Depression, stress and anxiety are the most common types of mental illness. It is common for people with these feelings to lack confidence and have low self-esteem despite having the same full range of intellectual abilities as the population as a whole.
- Maximise opportunities for success by assigning tasks that are neither too easy nor too difficult.
- Encourage individuals to find strategies that work best for them and enable them to become independent in their learning or work.
- Ensure that the individual knows who they should speak with within the organization for more support.

This is the inspiring story of a furniture store that learned to adapt. Widya Pertusi had a vision to share her Indonesian culture and craftsmanship with the American community in which she lived. Her original business was a furniture store that sold eclectic pieces of furniture from her native country to the residents of Summit, New Jersey.

The store wanted to enhance the shopping experience for their patrons, so in 2010 they began serving aromatic coffees. Customer response was so enthusiastic, that the store expanded into serving "fresh and healthy nouveau-inspired salads, create-your-own salad, Hale & Hearty soups, hot-pressed panini, gourmet sandwiches and wraps." This was a huge success and the customers began visiting regularly to indulge in the fresh offerings.

In an effort to adapt to consumer demand, they installed a full-service kitchen and expanded their menu even further. This led to a complete evolution in the business and birthed the Batavia Café'. Now, this full-service restaurant is serving delicious fare for breakfast, lunch, and dinner. They offer an array of healthy foods, which include daily vegetarian and gluten free soups. They have opened a catering business as well and are available for both small and large gatherings.

In a time when many businesses are experiencing a decline, it is inspiring to see a company that listened to the demands of their customer base and followed where it leads. This evolution from furniture store to café is unique and is sure to be told for years to come.

This is the best example and success story of Adaptability.

2.2.6 Positive Thinking:

"Try to offer anything happily and unconditionally if possible before you expect anything from anybody."

-LSS

The power of positive thinking is directly connected to the ability to remove, control or eliminate negative thoughts.

Watch your thoughts for they become words.
Watch your words for they become actions.

Watch your actions for they become habits.
Watch your habits for they become your character.
Watch your character for it becomes your destiny
Says Mahatma Gandhi.

When you see someone positive, you will be surprised to find such a positive energy spreading across even before you could start your interactions with such people. Such characters with bubbling energy are a boon to the humanity as they display positivity around them bringing great amount of easiness around and making people so comfortable. Anyone would like to talk to them with utmost comfort as they are the ambassadors of confidence and ease. This virtue is inborn as they are naturally so open, free to discuss, precise to the point, consoling, understanding etc. One who thinks positive only will have such virtue. This virtue can drive everyone around to positive thinking and bereft of fear of something going wrong.

With a positive attitude we experience pleasant and happy feelings. This brings brightness to the eyes, more energy, and happiness. Our whole being broadcasts good will, happiness and success. Even our health is affected in a beneficial way. We walk tall, our voice is more powerful, and our body language shows the way we feel.

Once a company manufacturing shoes sent their representative to an island country far away to assess the potential of selling their products there as a new venture there. Knowing this, their business competitor also sent their representative to the same place to assess faster and give a feedback urgently. The rep of the competitor company also landed there and both were roaming around and taking survey. The rep that came later was in a hurry to go back.

His observations were mailed to his headquarters like this:

"The population runs in thousands. The place is under developed and advent of modern civilization is not in the near future. The people are half-dressed only. Nobody has any knowledge of shoes or foot wear. I feel there is no point in our pushing our products here as there seems to be no market for it."

The representative of the first company, who landed first also sends a mail to his headquarters as below:

"The population runs in thousands. The place is under developed and advent of modern civilization is yet to begin. The people are half-dressed only. Nobody has any knowledge of shoes or foot wear. I feel there is plenty of scope in our pushing our products here as there seems to be great market for it. Send sample consignments urgently for trial launching."

I am sure one can understand the difference in the approach and attitude of both persons. A person, who approaches things in a positive way, has a larger scope to success if the implementation process is also done effectively.

This attitude keeps one growing with confidence which gets spread among all around. It brings in an assurance that there is always a solution for any problem as you are ready to cater to the need of the hour to think and act positive.

Your attitude of positivity will drive away the apprehensions others may have and will slowly pull them to your way of action which is good for the organization and the project.

Though you are a person of positivity, you must also realize that you cannot bring in miracles out of a hollow situation. You should also make people around you realize that all situations cannot be made good only by positive thinking, though it can be tolerated because of the anticipating character of the person having positive thinking.

As people around you have a great feeling about you, never get carried away and venture where you do not dare. By doing so, you may lose the confidence others have on you.

Positive thinking gives an inept feeling that your team members are going to willingly tow your line of thinking which will make everyone in the team to feel free to discuss, understand, cooperate and participate with full heart and soul.

Stop getting stuck in the past and focusing on when you "used to be good" or the mistakes you made back then. No matter if you think you were better then or worse, the only thing you should worry about is doing well in the future. You can't change the past but you can have a better future, so put all of your energy into that, instead of worrying about what you can't change.

Procrastinating is setting yourself up for failure. When you have less time to do a task, you'll rush and miss things. Instead, do things on time so that you have the extra time to really do your best!

Positivity is the biggest strength in life. Success follows who are positive in all situations. Many great celebrities had one or more weaknesses but they were always positive for self, others and situations and hence became successful by overcoming all sorts of obstacles. A little poison of negativity about others in our mind can destroy our relationships with them. Everybody has definitely something that can be appreciated, we only need to apply positive vision and appreciate others. Instead of thinking, "We are going to have a hard time adjusting to our present crisis situation," think, "We are facing some challenges in our present situation, but we will come up with solutions that we all will be happy with."

Here is a short story on positive thinking for the novice.

Two cockroaches lived in a house. Both of them were young and full of energy. Each day, they would run, jump and chase each other while playing. Though they were equally strong, there was a difference. One cockroach was optimistic and always lived in hope, while the other was pessimistic and lived in despair.

One day, while playing, both cockroaches fell into a pot of milk. They swam around and tried to hop out, but, as there was no solid support under their feet, it was not possible for the cockroaches to hop out and escape from the pot.

After some struggle, the pessimistic cockroach said. to itself, "It is impossible to hop out. No doubt, I have strength but I can't swim very long. I am already tired." Thinking thus, the cockroach did become tired soon and could not swim any longer. It gave up its struggle and went down to the bottom of the pot. Finally, it was drowned.

On the other hand, the optimistic cockroach kept on struggling, saying to itself, "No doubt, it seems difficult but who knows! Maybe some miracle will occur. If I try a little longer, something good might happen. It is a question of only a few minutes more and then I will be out of here."

Hoping for a miracle to happen, the second cockroach went on swimming. His constant leg movements churned the milk and turned it into a huge heap butter. Soon the cockroach was able to climb up the heap of butter and hopped out of the pot.

Positive thinking had saved the life of the cockroach.

In order to turn the mind toward the positive, some inner work is required, since attitude and thoughts do not change overnight.

1. Read about this subject, think about its benefits, and persuade yourself to try it. The power of your thoughts is a mighty power that is always shaping your life. This shaping is usually done subconsciously, but it is possible to make the process a conscious one. Even if the idea seems strange, give it a try. You have nothing to lose, but only to gain.

2. Ignore what other people say or think about you, if they discover that you are changing the way you think.

3. Use your imagination to visualize only favorable and beneficial situations.

4. Use positive words in your inner dialogues, or when talking with others.

5. Smile a little more, as this helps to think positively.

6. Once a negative thought enters your mind, you have to be aware of it, and endeavor to replace it with a constructive one. If the negative thought returns, replace it again with a positive one. It is as if there are two pictures in front of you, and you have to choose to look at one of them, and disregard the other. Persistence will eventually teach your mind to think positively, and to ignore negative thoughts.

7. In case you experience inner resistance and difficulties when replacing negative thoughts with positive ones, do not give up, but keep looking only at the beneficial, good and happy thoughts in your mind.

8. It doesn't matter what your circumstances are at the present moment. Think positively, expect only favorable results and situations, and circumstances will change accordingly. If you persevere, you will transform the way your mind thinks. It might take some time for the changes to take place, but eventually they will.

9. Another useful technique is the repetition of affirmations. This technique is similar to creative visualization, and can be used together with it.

Positive thinking paves way for people to recognize you sooner as a potential leader.

2.2.7 Self Discipline:

> *"Do not take anybody for granted. They have their own identity and they need recognition from the society like you do."*

>   -LSS

An American Indian story goes like this.

An old Cherokee was explaining the modus operandi of life to his son. He said that a fight was always going on inside him like any other individual in the universe. The boy asked what it was.

The Cherokee explained that two wolves were fighting inside him. One was "Evil" and the other was "Self Discipline".

The Evil was fighting with ammunitions like anger, greed, arrogance, guilt, false ego, lies, inferiority and all vices while the other wolf Self Discipline was fighting with its ammunitions like goodness, peace, happiness, love, kindness, humility, empathy, benevolence, truthfulness etc.

The boy asked him as to which wolf won the fight.

Cherokee snapped, "The one any individual was feeding!"

So, it is clear that it is only our own choice to be good or bad.

Self-Discipline is the behavior of being sensible, straight, honest and true to yourself and the organization you belong to. There is no need for false showing and deceptiveness.

Talent without discipline is like an octopus on roller skates. There's plenty of movement, but you never know if it's going to be forward, backwards, or sideways.

Self discipline is not a new idea. There is an old story about a man who went to a tattooist because he had always wanted a tattoo of a lion on his back.

The tattooist started to sketch the tail into the man's torso: 'Ouch! What are you doing?' asked the man. 'I'm doing the lion's tail' replied the tattooist. 'Well then for goodness sake let's have a lion without a tail!' said the man, wincing in pain.

Next the artist set about on the Lion's whiskers. 'Ouch!' cried the man, 'What's that?' 'The whiskers!' said the tatooist, getting increasingly irritated. 'Well let's have a lion without whiskers!' moaned his customer.

The tattooist then set about doing the Lion's back. 'No that hurts too!' shouted the man. At this, the tattooist finally lost patience with the man's lack of self discipline. Throwing down his tools and the man out of his shop he shouted, 'How can you expect to get what you want without a little discomfort?'

One meaning of this story may be to show how handicapped you are if you base your decisions *purely* on your comfort level. If we don't develop the capacity for self discipline we deprive ourselves of not only greater likelihood of success, but also larger and lasting satisfactions.

Knowing we can discipline ourselves over and above what feels comfortable increases self confidence. We need to be stretched as much as we need comfort and rest.

"Don't have a wishbone where your backbone should be!"

Self-discipline can be considered a type of selective training, creating new habits of thought, action, and speech toward improving yourself and reaching goals.

Self-discipline can also be task oriented and selective.

Instead of devoting a lot of hours one day, and none the other and then a few on an another day and so on, allocate a specific time period each day of the week for that task and ensure that you are holding on to it.

When you first begin your work day, or going to work take a few minutes and write down on a piece of paper the tasks that you want to accomplish for that day.

Prioritize the list.

Immediately start working on the most important one.

Try it for a few days to see if the habit works for you.

Habits form over time: how much time depends on you and the habit?

Observe the people in your life and see to what extent self discipline and habits help them accomplish goals. Ask them for advice on what works, what does not.

Self-discipline involves acting according to what you think instead of how you feel in the moment. Often it involves sacrificing the pleasure and thrill of the moment for what matters most in life. Therefore it is self-discipline that drives you to work on an idea or project after the initial rush of enthusiasm has faded away.

Discipline means behaving according to what you have decided is best, regardless of how you feel in the moment. Therefore the first trait of discipline is self-knowledge. You need to decide what behavior best reflects your goals and values. This process requires introspection and self-analysis.

As you begin to build self-discipline, you may catch yourself being in the act of being undisciplined – e.g. biting your nails, avoiding the gym, eating a piece of cake or checking your email constantly. Developing self-discipline takes time, and the key here is you are aware of your undisciplined behavior. With time this awareness will come earlier, meaning rather than catching yourself in the act of being undisciplined you will have awareness **before** you act in this way. This gives you the opportunity to make a decision that is in better alignment with your goals and values. Self-talk is often harmful, but it can also be extremely beneficial if you have control of it. When you find yourself being tested, I suggest you talk to yourself, encourage yourself and reassure yourself. After all, it is self-talk that has the ability to remind you of your goals, call up courage, reinforce your commitment and keep you conscious of the task at hand. When I find my discipline being tested, I always recall the following quote: "The price of discipline is always less than the pain of regret". Burn this quote into your memory, and recall in whenever you find yourself being tested. It may change your life.

The five pillars of self-discipline are: Acceptance, Willpower, Hard Work, Industry, and Persistence. If you take the first letter of each word, you get the acronym "A WHIP" — a convenient way to remember them, since many people associate self-discipline with whipping themselves into shape.

When we talk of hard work, I remember the proverb from the Holy Bible. The Scriptures spell it out: *"In all labour there is profit: but the talk of the lips tendeth only to penury [poverty]" (Prov. 14:23). "He that tilleth his land shall have plenty of bread: but he that followeth after vain persons shall have poverty enough" (28:19).*

In Chapter 16 of the Bagwat Gita Sri Krishna elaborates on two types of Work Ethic viz. daivisampat or divine work culture and asurisampat or demonic work culture.

Daivi work culture - means fearlessness, purity, self-control, sacrifice, straightforwardness, self-denial, calmness, absence of fault-finding, absence of greed, gentleness, modesty, absence of envy and pride.

Asuri work culture - means egoism, delusion, desire-centric, improper performance, work which is not oriented towards service. It is to be noted that mere work ethic is not enough in as much as a hardened criminal has also a very good work culture. What is needed is a work ethic conditioned by ethics in work.

Don't compare yourself to other people. It won't help. You'll only find what you expect to find. If you think you're weak, everyone else will seem stronger. If you think you're strong, everyone else will seem weaker. There's no point in doing this. Simply look at where you are now, and aim to get better as you go forward.

A teacher and his student were walking from one village to another, when they suddenly heard a roar behind them. Turning their gaze in the direction of the roar, they saw a big tiger following them.

The first thing the student wanted to do was to run away, but since he has been studying and practicing self-discipline, he was able to halt himself from running, and wait to see what his teacher would do.

"What shall we do Master?" Asked the student.

The teacher looked at the student and answered in a calm voice: "There are several options. We can fill our minds with paralyzing fear so that we cannot move, and let the tiger do with us whatever pleases it. We can faint. We can run away, but then it will run after us. We can fight with it, but physically it is stronger than us."

"We can pray to god to save us. We can choose to influence the tiger with the power of our mind, if our concentration is strong enough. We can send it love. We can also concentrate and meditate on our inner power, and on the fact that we are one with the entire universe, including the tiger, and in this way influence its soul."

"Which option do you choose?"

"You are the Master. You tell me what to do. We don't have much time." The student responded.

The master turned his gaze fearlessly towards the tiger, emptied his mind from all thoughts, and entered a deep state of meditation. In his consciousness, he embraced everything in the universe, including the tiger. In this state the consciousness of the teacher became one with consciousness of the tiger.

Meanwhile the student started to shiver with fear, as the tiger was already quite close, ready to make a leap at them. He was amazed at how his teacher could stay so calm and detached in the face of danger.

Meanwhile the teacher continued to meditate without fear. After a little while, the tiger gradually lowered its head and tail and went away.

The student asked his teacher in astonishment, "What did you do?"

"Nothing. I just cleared all thoughts from my mind and united myself in spirit with the tiger. We became united in peace on the spiritual level. The tiger sensed the inner calmness, peace, and unity and felt no threat or need to express violence, and so walked away."

"When the mind is silent and calm, its peace is automatically transmitted to everything and everyone around, influencing them deeply", concluded the teacher.

Write or record any thoughts or negative feelings you have about yourself, then delete or destroy them later. You'll be amazed how great you feel if you let a few things out! Do it often to build up self-esteem.

Do not compare yourself with others. That should not be the basis.

You have to know that you are so much better than what they say...

You have to do what you want because there's no one who can stand in your place.

Achieving a goal, however small, will build self-esteem.

After all the discussions we had on the qualities of a novice, I was glad that I did not fall short of the required qualities to improve myself to the next higher level of Middle Management. Now, let us see and explore Middle Management.

SECTION 3.0

MIDDLE LEVEL MANAGEMENT

3.0 Middle Level Management

"You may doubt someone for some time for some reasons, but not all the time for all his deeds. You may ultimately loose him forever."

-LSS

A novice having imbibed most of the desired qualities discussed in previous chapters and having involved himself thoroughly in work with all sincerity for a few years and gathered the practical hands on experience for three to four years, gears up herself up to the Middle Level Management. Her confidence level increased to a larger extent, unafraid of challenges. She has lots of ideas in circumstances of difficulty. Her qualities improve to a finer and wider perspective for the good and towards the path of taking him to higher levels. The aspects which she adds and gains at this level can be categorized as follows:

3.1 Maximizing ability
3.2 Consistency
3.3 Analytical Ability
3.4 Build Relationship
3.5 Self-Assuring.

I had some reservations on management by Engineers earlier as there was no subject on management taught to them in their regular courses. Nevertheless, there are several examples of Engineers being very good managers. But the fact is that those Managers from Business schools do feel differently on the capability of Engineers as Managers.

A man in a hot air balloon realized he was lost. He reduced altitude and spotted a woman below. He descended a bit more and shouted, "Excuse me, can you help me? I promised a friend I would meet him an hour ago, but I don't know where I am."

The woman below replied, "You're in a hot air balloon hovering approximately 30 feet above the ground. You're between 40 and 41 degrees north latitude and between 59 and 60 degrees west longitude."

"You must be an engineer," said the balloonist. "I am," replied the woman, "How did you know?"

"Well," answered the balloonist, "everything you told me is, technically correct, but I've no idea what to make of your information, and the fact is I'm still lost. Frankly, you've not been much help at all. If anything, you've delayed my trip."

The woman below responded, "You must be in Management." "I am," replied the balloonist, "but how did you know?"

"Well," said the woman, "you don't know where you are or where you're going. You have risen to where you are due to a large quantity of hot air. You made a promise which you've no idea how to keep, and you expect people beneath you to solve your problems. The fact is you are in exactly the same position you were in before we met, but now, somehow, it's my fault."

3.1 Maximize Ability:

> **"Whatever is flexible and flowing will tend to grow. Whatever is rigid and blocked will wither and die. So are your loved ones."**
>
> **-LSS**

> "The difference between what you were yesterday, and what you will be tomorrow, is what you do today..." ~Stephen Pierce

Failing to plan is planning to fail. Have an end goal in mind and organize a game plan to get there. Create a daily schedule and stick to it. Be punctual and don't postpone. Determine what projects and tasks are critical, urgent and create the largest value and benefit to the business. Select what you can easily reprioritize and what you can delegate without hurting results; a lion's chunk of prioritizing involves simply learning to say 'No'.

What if I did this differently? What if there is a faster, easier, more effective way to achieve the same results? What if I changed my pitch/tone/content/brand or message? Whatever the question you ask, you need to constantly challenge the status quo and question yourself and others to find out whether you are really working as productively as you possibly could. Never be satisfied that things can't be improved.

Passion is one of the single most important drivers of success. If your heart and soul aren't in what you are doing, you are likely not putting your best effort forward. Do some soul-searching and find out what it will take to make you love what you do and perform to the very best of your abilities. Make sure your values are being reflected and respected in what you do and that you see true meaning in your work.

Effective and timely feedback is a critical component of a successful performance management program and should be used in conjunction with setting performance goals. If effective feedback is given to employees on their progress towards their goals, employee performance will improve. People need to know in a timely manner how they're doing, what's working, and what's not. Feedback can come from many different sources: managers and supervisors, measurement systems, peers, and customers just to name a few. However feedback occurs, certain elements are needed to ensure its effectiveness.

Feedback should be given in a manner that will best help improve performance. Since people respond better to information presented in a positive way, feedback should be expressed in a positive manner. This is not to say that information should be sugar-coated. It must be accurate, factual, and complete. When presented, however, feedback is more effective when it reinforces what the employee did right and then identifies what needs to be done in the future. Constant criticism eventually will fall upon deaf ears.

> *"The obstacles in our path ARE the path. Every time we stretch beyond our resistance and our fear, we make a choice for life. And, every time we choose life, we find that fear loses its grip on us..." ~ Rolf Gates*

3.2 Consistency:

"Mind is a treasure box to keep love, happiness and sweet memories, not a garbage bin to nurture anger, hatred and jealousy!"

-LSS

Let's say, your dearest friend needed a favor and you immediately promised him that you will do it. Days went by and you kept procrastinating. Finally you forget about it. Then the time comes, when you apologize to him. He is ok with it, but you are not. You feel miserable inside. Your guilt eats you from the inside. Finally to help yourself, you try to find out reasons to justify yourself. You rationalize that you had important personal things going on, maybe because it was not worth all the trouble, and maybe because he should learn to do his own stuff. In your mind, you know that you are being unreasonable. You know that you could have, and you should have kept your promise. This kind of post-rationalization behavior to ease our mental tensions is called the Consistency theory.

Usually when our thoughts, attitude, beliefs all our in agreement with each other and things around us also support these thoughts then our mind is at peace and comfort. Cognitive dissonance is what can cause unrest in our minds. This is when our mental peace fails, and things seem in disorder and there is turmoil in general. This is when the consistency theory kicks in. Our human mentality tries to find a comfortable situation to the ongoing problems. We try to achieve a balanced state which gives us some mental peace.

A lot of research has been carried out on this theory. People try to find out ways in which we try to achieve this mental state. These are:

1. To deny or completely ignore the happening of the incident. (E.g. I was never told about this)

2. To find a rational explanation or excuse. (E.g. That would have happened anyways)
3. By separating the item. (E.g. While defending pollution, an argument like I don't think my single car is going to make much of a difference in this whole lot of pollution generated)
4. By being transcendent. (E.g. There are others doing it too. So it's not just my fault. There are so many others erring too)
5. By agreeing to improve. (E.g. In the future I will do it on time)
6. To persuade our self that it never was our mistake and what we did was in fact right.

In any team sport, the best teams have consistency and chemistry. And so is the case with business.

Business growth requires a track record of success. You can't establish a track record if you are constantly shifting gears or trying new tactics. Many efforts fail before they get to the finish line, but not because the tactic was flawed or goals weren't clear. The problem is often that the team simply didn't stay the course to achieve the objective.

Your employees and your customers need a predictable flow of information from you. All too often I see businesses, both small and large, adopt a campaign or initiative only to end it before it gains traction. It's effective to run many advertisements, numerous blog entries, weekly newsletters, or continual process changes throughout a year.

Your team pays as much or more attention to what you do as to what you say. Consistency in your leadership serves as a model for how they will behave. If you treat a meeting as unimportant, don't be surprised when you find they are doing the same to fellow teammates or even customers.

When our inner systems (beliefs, attitudes, values, etc.) all support one another and when these are also supported by external evidence, then we have a comfortable state of affairs. The discomfort of cognitive dissonance occurs when things fall out of alignment, which leads us to try to achieve a maximum practical level of consistency in our world.

Highlight where people are acting inconsistently with beliefs, etc. that support your arguments. Show how what you want is consistent with the other person's inner systems and social norms.

This is a very popular story to highlight the importance of consistency in whatever you do.

Once there were two friends - a hare and a tortoise. The hare was known for his swiftness and the tortoise was known for his sluggishness. The tortoise was extremely slow.

One day, as they chatted, the hare began making fun of the tortoise for his slowness.

The tortoise was slightly annoyed but said with a smile, "I may be slow, but I can beat you in a race."

The hare was astonished to hear this. He thought the tortoise was utterly foolish and totally unaware of what he could not do, even in his wildest dreams.

"Are you kidding?", said the hare in bewilderment. "I hope you are not serious."

"I am very serious. I am sure I can outrun you," said the tortoise. Seeing the tortoise so serious, the hare said, "All right, in that case, we shall appoint a referee and fix a venue for the event."

On that note they parted, to meet again on the appointed date.

A rat was appointed as referee. A large field beside the river was selected for the unusual race and a big banyan tree, about a mile away from the rat's hole, was decided on to be the winning post.

The rat stood ready to blow the whistle and start the race. The tortoise and the hare tensed at the start line.

"On your mark, get set, go", called the rat, and the race began.

The hare took off at lightning speed, and soon ran out of sight towards the finish line. Meanwhile the tortoise began the race at a very slow pace. The sight was almost funny if not pitiful.

"Poor tortoise," thought the rat, "The hare will win the race hands down, and cover the length of the field ten times before the tortoise can cover it even once. No match at all!"

The hare must have reached about half a mile when he stopped to see where the tortoise was. He looked back. The tortoise was not to be seen. "Oh he is far behind; I can't even see him yet. I think I will wait here until I can see him and then I'll run the remaining distance. Hey, why don't I eat some grass and rest in the meanwhile." said the hare to himself.

The hare snacked and drank some water, and lay down in the shade of a tree to wait and watch. Soon the cool air from the riverside lulled him into deep sleep. The tortoise on the other hand, kept moving slowly but steadily

The hare slept for a long time. When he woke up, he looked around and the tortoise was not to be seen anywhere. He felt rested and so decided to complete the race. As he approached the finish line, he grew more and more astonished. The tortoise had already reached the finish line.

The hare had lost the race. He accepted the defeat graciously. After that he never poked fun at the tortoise or his slowness.

The moral of the story: If you have all that you need to win the race, the only thing that could stop you from winning the race is lack of persistence in effort.

Which one of the two was Consistent? The tortoise of course. What are the characteristics of Consistency?

- He believed that it did not matter how tough the goal was, if he kept at it and did a little every moment, he would be able to achieve it.
- It did not matter what others were able to do or not able to do, goal achievement meant, that he should do his piece consistently.
- To win he needs to put in work steadily even if it is slow.
- He took time to do all the right things at the right time.
- Doing what you need to do slowly, steadily day in and day out makes goal achievement effortless.

What are the non-attributes of Consistency that made the hare lose the race?

- Ability is inadequate, if effort is inconsistent.
- He knew he could do it ten times over and well. Yet he thought that the effort could wait till the goal was near. He left putting in the effort till too late.

Hence consistency is the key to success in whatever you do.

3.3 Analytical Skill:

"Happiness cannot be travelled to, owned, earned, worn or consumed. It is the spiritual experience of living every minute with love, grace and gratitude."

There's been a growing focus on analytical skills and people's ability to find solutions for problems of advanced complexity. So much emphasis has been given to analytical skills - such as critical thinking, problem solving and information-to-knowledge conversion - that people don't pay adequate attention to improving their communication skills. In fact, communication skills, such as the ability to confidently and clearly express an opinion or being convincing, are often considered secondary to allegedly more important skills.

A methodical approach to analyzing business initiatives is key to organizational planning and decision making. A critical ingredient in the evolving role of today's project manager is a solid understanding of the principles, methods, tools and best practices for evaluating proposed projects for the enterprise.

The ability to make high-impact, well-considered decisions under less-than-ideal conditions, as well as strong leadership, communication and problem-solving skills, are critical for sustaining competitive advantage.

Analytical skill is the ability to visualize, articulate, and solve both complex and uncomplicated problems and concepts and make decisions that are sensible and based on available information. Such skills include demonstration of the ability to apply logical thinking to gathering and analyzing information, designing and testing solutions to problems, and formulating plans.

People with strong analytical skills can see and discuss issues, make decisions, and solve problems, even under difficult circumstances. It is important that they follow the ethics of business with wisdom and righteousness. *This is what the Eight Fold Paths of Buddha says, which are Right View, Right Intention (both representing Wisdom), Right Speech, Right Action, Right Livelihood (representing Ethics), Right Effort, Right Mindfulness and Right Concentration (representing Samaadhi). Wisdom makes us to wonder*

"Why", Ethics make us think "How" and Samaadhi makes us think "What". This analytical management has been in existence for a long as seen from preaching of Lord Buddha.

Having an analytical mindset does have its setbacks; People may place too much trust in data or may not trust the data unless they have had direct experience to support it.

The sole responsibility of an Analytical Manager is to configure, design, implement and support of the data analysis solution or business intelligence tool. This area of analytics is gaining more and more importance and popularity in the industry of information technology which is a sub area of statistics.

The role of analytical managers is like diving into the sea and collect natural pearls from inside the sea shells. It is very hard to identify from the sea bed as to which one is the clam shell containing pearls and which one is sea animal or under-water insects. In fact, they have to deal effectively among contradictions, illogical ideas and information they get. They have to translate illogical and contradictions into an acceptable pattern which is useful for the organizational growth. This is not that easy. They need to concentrate sensibly and they prefer to work alone for formulating a reasonable theory acceptable to all.

If you were to test for analytical skills you might be asked to take a series of events and put them in proper sequence, look for advertisement inconsistencies, or read an essay with a critical eye. There are generally standardized tests used as a guideline. In all instances analytical skills require you to dissect a problem and then find a solution for that problem. There are many skills required within a company and its workforce, but analytical skills are essential in every organization.

To test if someone has analytical skills, he might be asked to look for inconsistencies in an advertisement, put a series of events in the proper order, or critically read an essay. Usually, standardized tests and interviews include an analytical section that requires the examiner to use their logic to pick apart a problem and come up with a solution.

Although there is no question that analytical skills are essential, other skills are equally required as well. For instance, in systems analysis, the systems analyst should focus on four sets of analytical skills: systems thinking, organizational knowledge, problem identification, and problem analyzing and solving.

It also includes the way we describe a problem and, subsequently, find the solutions to these problems. The more senior a manager you are, the less contact you have with the day-to-day realities of your organization and the more you have to rely on information that is provided to you by others. This information is always subject to some sort of filter based on the assumptions you make and the questions you ask. In the analytical mindset, you are focused on making decisions based on imperfect and incomplete information.

The downsides of an analytical mindset are either that you place too much trust in the data, not realizing that it is flawed, or that you mistrust any data unless you have had direct experience.

Someone working well in this mindset uses what information they have judiciously and is prepared to cope with a certain level of ambiguity or uncertainty.

In career terms, someone who feels that they cannot evaluate a career option until they have experienced it or who assumes they know everything about an option without researching it, may be suffering from problems with their analytical mindset.

Analytical skills and critical thinking are used to break down, assess and evaluate problems or concepts in order to make decisions or come to conclusions that are based on the synthesis of available information.

Essentially **analytical** skills are the main requirement along with the ability to understand and contextualize the business environment.

Analytical thinking is a critical component of visual thinking that gives one the ability to solve problems quickly and effectively. It involves a methodical step-by-step approach to thinking that allows you to break down complex problems into single and manageable components.

Analytical thinking involves the process of gathering relevant information and identifying key issues related to this information. This type of thinking also requires you to compare sets of data from different sources; identify possible cause and effect patterns, and draw appropriate conclusions from these datasets in order to arrive at appropriate solutions.

A manager's analytical skills are frequently used to solve problems. Much like a detective, a manager is often called upon to solve business problems. When units are not performing as expected or a crisis develops in an important location, managers must use their analytical skills, especially questioning and researching, to determine what is actually causing the problem. Once a problem is analyzed and a potential cause is determined, a manager must select the appropriate solution to implement. Problem-solving skills are an important tool for the analytical manager.

To analyze is to break larger concepts into smaller parts. Primary analytical skills, therefore, are those that demonstrate the ability to develop a clear line of thinking based on logic and to reach a sound conclusion based on that logic. When describing analytical skills, it is helpful to use terms that apply appropriately to qualitative and quantitative analysis. Qualitative analytical skills are those that deal more with abstract concepts.

Crime solving a fitting example of qualitative analysis. Detectives In order to solve a crime, must analyze many different types of evidence which or relevant or irrelevant to the case. Tying the less obvious facts to the obvious requires refined analytical skills. Quantitative analytical skills are those that pertain to specific numbers and data. Similar is the application of this analysis in companies and organizations to pin point the profit and loss areas.

"The modern mind tends to be more and more critical and analytical in spirit, hence it must devise for itself an engine of expression which is logically defensible at every point and which tends to correspond to the rigorous spirit of modern science" says Edward Sapir.

Ruchi Sanghvi describes this aspect as, "Engineers love to optimize problems. Now I optimize logistical problems. I ask: 'What's the goal? What are our constraints? What is the optimal, elegant way to get to that goal within those constraints?' I break it down in terms of a data funnel: 'Where in the funnel are we inefficient?' That analytical background really helps."

3.4 Build Relationship:

> *"Positive thoughts generate positive feelings and attract positive life experiences."*

> *-L.S.S*

It's not always clear to employees why they need to manage relationships upward unless it's for political maneuvering or brown-nosing. But it is a valuable skill to know how to consciously work with your boss to obtain the best possible results for you, your boss, and the organization you both work for.

This is important because you and your boss (and the extended team) are ***mutually dependent*** on one another. Your boss and team need your help and cooperation to deliver effectively and you need your boss' and team's support and guidance in doing your job effectively.

We often make either the mistake of seeing ourselves as not very dependent on our colleagues or assuming that our bosses will magically know information or help we need from them without asking for it.

The keys to managing our bosses effectively are:

- Have a good understanding of your boss and of yourself – each of your strengths, weaknesses, work styles, goals and pressures.
- Use this information to develop and manage a healthy working relationship – one that is compatible with people's work styles and strengths, where expectations are mutual and shared, and where each person's most critical needs are met.

It is important to understand your boss - not just initially when you first begin working with one another, but throughout your relationship. On an ongoing basis, it's important to communicate as priorities and concerns change.

It is even more important to know yourself and your strengths, weaknesses, work style preferences, goals, and pressures. Developing self-awareness and then learning to apply your knowledge in order to have more effective relationships is an important ongoing learning process that we all need to and should engage in.

Learning more about your own and others' preferences for how they take in information, make decisions, structure their day, and communicate is something that can serve as a foundation for understanding work styles and strengths.

Find ways to regularly communicate your expectations to your boss, receive feedback and ask questions about his/her expectations, and influence him/her to your point of view on important issues. But realize that your boss, just like you, is probably limited in his/her time and energy, so make sure that you use his/her time wisely.

One of the interesting things that we see in business meetings and networking events is how, in a room full of people, you can successfully exclude others from joining in sometimes without realizing! When people try to monopolize the time of others, it precludes anyone else joining in and this is contrary to good relationship-building. Be sure to check that you are not the one doing the monopolizing otherwise you become the party bore that no one will want to talk to.

With the global nature of our business, employees often come from different countries, each with a different culture. In order to successfully integrate multicultural differences, these differences must be understood, articulated, and respected.

Consultations not only benefit the initiator but the entire team involved in it. The importance of consultation with subordinates can be seen in Ramayana.

Rama Rajya is the benchmark of good governance, so it is worthwhile to recapitulate the golden words of wisdom from *Valmiki Ramayana*.

The second chapter, the 'Ayodhya Kanda', offers comprehensive lessons on good governance. Bharata, the younger brother of Rama, goes to meet him in the forest to request him to return to Ayodhya and rule. Rama declines and counsels Bharata on governance. From the quality of ministers and the importance of strategy sessions, to temperance in administration and justice, he expounds all the subtleties of statecraft in a lucid manner. Though Rama seems to be enquiring about Bharata's well-being and whether all is well at Ayodhya, he is, in fact, offering lessons on effective governance. The dialogue between the two brothers runs into several pages, but some important lessons are obvious.

Emphasising the quality of ministers, Rama asks Bharata whether he has appointed courageous, knowledgeable, strong-willed men with high emotional quotient as his ministers, because quality advice is the key to effective governance. The emphasis is on competence and confidentiality. Rama's advice to Bharata is to take a decision on a complex issue neither unilaterally nor in consultation with too many people.

Rama also advises Bharata to prefer one wise man to a thousand fools as it is the wise who can ensure prosperity during an economic crisis. Appointing tested men of noble lineage and integrity for strategic positions is the key to successful Government. Moderate taxes should be levied on people, lest they revolt. Rama wants Bharata to treat his soldiers well and pay legitimate wages on time.

Protecting forests and maintaining livestock have also been dealt with as important aspects of effective governance. In fact, the vision of the *Ramayana* has eternal relevance. Law and justice, finance and business, corruption and injustice to the poor are all mentioned. Rama's advice to Bharata is as relevant today as it was in the Treta period.

3.5 My Experiences As Middle Level Manager

> *"Whatever you do, good or bad, people will always have something negative to say. Practice expecting this and to ignore once."*

> *-LSS*

Having seen the details of the prime characteristics of Middle Level Manager and having understood the same, let me refresh my memory of having experienced such occasions which are ever green to me as a successful middle level officer in government service. This level of management is a critical path one has to tread in carefully as he or she has to manage both levels,- the level above as well as the level below. You have to be a hero to your staff and subordinates while at the same time, a faithful and trusted staff to your boss. Sometimes the middle level managers tend to think that they are only right and get annoyed with the seniors. This is because that the thinking of these persons are one dimensional and may not be aware of the fact that there are some more angles to the situation as they might not have had the opportunity to know them and those angles were not needed to be communicated to them.

Hence, it is always good for this middle level managers to participate in the brain storming sessions if they get any opportunity but should not worry if their ideas were not considered while decisions are made. The nuances of this would be understood by them only when they grow more in the hierarchy and reach higher levels. I had several such experiences but soon reconciled to reality.

3.5.1 My First Acid Test:

On promotion as Executive Engineer (E.E.)and getting posted at Kavaratti, the capital of Lakshadweep group of islands, I landed in Calicut (Kozhikode) in January of 1986 and reported to my boss. After preliminary rounds of talk and formal introduction to our staff, I sat in a chamber next to that of my boss. He was in a meeting with two people and soon he called me to his chamber. He introduced me to them as new EE and them as representative of Cochin Marine Corporation, Cochin (CMC) who had completed construction of a Tug (high powered boat used in ports for pulling, pushing, towing of ships and floating crafts; in short a scaled down version of a ship) to A.L.H.W against a contract signed by my organization and the CMC. They could not settle the finances for a long time due to variances in delivered quantities against the agreement. My boss grinned at me and said, "Sekar, please hear their problems and see how we can resolve this issue. After all, this is the responsibility of the EE."

Unaware that no government officer would risk such a task as one of the first assignments coupled with the fact that I genuinely wanted to help, I invited them to my office. I also requested the concerned dealing hand, on the subject matter to join us with all relevant records. I started, "Gentlemen, this is my first day as EE and this is the first assignment I am dealing. I do not know the back ground of the case. So I want to hear the real facts from you. Please go ahead."

Another gentleman who represented CMC started explaining supported by a colleague from his company. I told my representative to interfere if any of their submission was not true. They started explaining their side of the story. "Sir, Initially, the items and quantities of the estimate were based on design and drawings given by a retired Merchandile Marine Surveyor (MMD Surveyor) which was approved by the then sitting MMD Surveyor and subsequently by your organization. Accordingly, estimates and tenders were called for and we won the order for execution.

During the course of execution, there were changes of officers in MMD office at Cochin and during every round of inspection by MMD Surveyor at different stages of fabrication, certain changes like deletions or additions were suggested which has led to a situation of variations in the quantities of the different items in the schedule of the agreement. The officer who is to pass our bill is objecting to variations and demands for the approval of competent authority. From his point of view he may be correct too. But the problem is we are held up for payment for no fault of us."

I asked my dealing hand, "Who is the competent authority?"

"Our boss Sir" came the response.

"Why was it not put up to boss for approval?"

"He wants justification on file from EE, Sir"

"Of course, he will ask for it. But why was justification not given?"

"Sir, We do not know how to justify."

I understood the undercurrent. Being a government body, there was no willingness on anybody's part to take any risk for fear of being blamed if things go wrong. But I was determined to go ahead and crack the issue as this was my first assignment.

I said "Tell me, do we have a tabulation of item-wise quantities as per agreement and the actual quantities executed against each?"

"Yes Sir"

"Good. Do we have any record on the changes suggested by MMD Surveyor during each inspection?"

"Yes Sir"

"That is great. Then there should not be problem anymore."

I could see their troubled face express more confusion. I could hear them thinking to themselves "When this dynamic boss (as he was) could not solve this problem, how is this new little chap going to solve this!?"

I smiled to myself and started dictating to my stenographer with a cautionary note to vendor representatives that they could stop me if I went wrong anywhere or if they had any objections on my understanding and observations.

I started dealing item wise and asked for variations for each, supporting document in the form of inspection notes of MMD Surveyor which could vouch for the variations. By applying the changes suggested by MMD, if the variations in quantity was tallying with their claim, I put it on record as "The variation in quantity of this item is justified based on the recommendations of MMD vide references dated ... who is the authority to inspect and order changes."

If the quantity executed as excess in some items fell beyond the recommendations of the MMD, I, mercilessly, disallowed the excess quantity beyond recommended quantity and recorded as "This excess quantity is disallowed for payment as it is unauthorized."

Immediately, one of them got startled and said "Sir, We have spent out finances on it. How can you reject it? We will suffer greatly. Kindly consider."

I shot back, "You said this problem is lying unsettled for more than two years. I want to settle this once and for all. You can't expect that you will be paid fully as per your claims. I am not disallowing anything which is required to be executed as per records. I am strictly going based on records. If you do not accept my decisions, probably you may have to wait for few more years. The choice is yours!"

Immediately, his other colleague who realized the logic behind my argument intervened. "Sir, so far nobody took initiative to touch this issue. Your approach to the problem is very logical and fair. So please go ahead and finalize. I feel confident that we will able to resolve this matter with your leadership."

As I continued to finalize my recommendations, I sensed the persisting discomfort in with the vendor. He wanted to interject at many occasions but was hesitant because of what I had just told him. I acknowledged his discomfort with a smile and a pat on his shoulder and closed my dictation with the words, "These recommendations are based on the records made available at the Circle office (Office of my boss). In case of any claims pending due to or from the agency is found from the division or sub division offices, they are also to be taken into account before release of final payment. Hence a part payment as per the quantities allowed as above can be approved for release."

With these words, I saw the expression on their faces change from one of frustration to and elation. As a dealing hand, he knew that there may be certain minor things to be sorted out from site and I too knew it because I too rose from bottom only.

After half an hour, my boss called me and said, "Mr. Sekar, They were having high praise on you for the way you handled the case. Very good! Keep it up!"

I never realized that my boss was going to be critical when someone appreciates his subordinate whom he had never acknowledged. There were many problems that cropped up when there were further appreciations on my performance. All those situations paved way for me to understand work place politics where I was still a novice! My boss was the kind of person that expects himself to be the center of the office. He wanted to be the one and only one who knows best and expects everyone to believe it.

A sales rep, an administration clerk and the manager are walking to lunch when they find an antique oil lamp. They rub it and a Genie comes out in a puff of smoke. The Genie says, "I usually only grant three wishes, so I'll give each of you one wish each."

"Me first! Me first!" says the admin clerk. "I want to be in the Bahamas, driving a speedboat, without a care in the world."

Poof! She's gone.

In astonishment, "Me next! Me next!" says the sales rep. "I want to be in Hawaii, relaxing on the beach with my personal masseuse, an endless supply of Pina Coladas and the love of my life."

Poof! He's gone. "OK, you're up," the Genie says to the manager.

The manager says, "I want those two back in the office after lunch."

Most bosses out there are great people. However, always let your boss have the first say. And then supplement the discussion with your thoughts in a polite and courteous manner. I took this lesson with me throughout my career. This is an essential lesson, particularly for those who are serving in Government organizations.

With happiness of cracking my first assignment most successfully, I moved to Kavaratti, the capital of Lakshadweep group of Islands which was the Headquarters of Executive Engineer. Before I go to my next episode of events, let me give a brief of Lakshadweep.

Lakshadweep, the group of 36 islands is known for its exotic and sun-kissed beaches and lush green landscape. The name Lakshadweep in Malayalam and Sanskrit means 'a hundred thousand islands'.

Lakshadweep is the tiniest Union Territory of India and is its only coral island chain. This archipelago consists of 36 islands, 12 atolls, 3 reefs and 5 submerged banks. The islands have a total area of 32 sq.kms and the lagoons enclosed by the atolls cover an area of 4200 sq.kms. Its territorial waters extend to 20,000 sq.kms and Exclusive Economic Zone (EEZ) to 4,00,000 sq.kms. Only 10 of these islands namely, Agatti, Amini, Andrott, Bitra, Chetlat, Kadmat, Kalpeni, Kavaratti, Kiltan and Minicoy are inhabited. Kavaratti is the Administrative Headquarters of the Union Territory. The islands are restricted area and permit from the Administration is required to visit the islands.

The population of Lakshadweep was only 64,429 (As per 2011 Census). The head of Administration is called Administrator assisted by Collector cum Development Commissioner, both from Indian Administrative Service.

Bitra is the smallest island with hardly 0.1 Sq.Km of land area and having a very vast area of charming blue lagoon all around it. There is no water resource and the population of this island was hardly about 100. Electricity was produced from solar panels only and power was supplied to the population for a few hours only during night time. Water was produced by evaporation and condensation of the sea water and so very scarce. Water was supplied in rationing to the population. As the available land area was very small the same was very precious that they cannot afford to use even a very small portion of land due to sea erosion. But unfortunately the nature was acting otherwise. There was severe erosion due to wave action and this was to be stopped on top priority. Hence we undertook shore protection works by getting stone boulders from Mangalore. We had mobilized about twenty workmen and me along with one Mr.Kanakadasan, then Junior Engineer and started the work. You can imagine that our team of about twenty odd persons definitely could have hampered their water rationing (as we were constituting about 20 to 25% of the existing population. But the people were so happy to accommodate us and extended all help required for our stay and execution of work. Mr.Kanakadasan stayed there for more than a month and completed the work successfully to the appreciation of all braving the adverse conditions prevailed over there. Kudos Kanakadasan!

3.5.2 Great Escape – Planning, Design and Execution of Break water, Andrott:

"When people say they cannot see anything good in you, hug them. Because Life can be very difficult for the blind."

Let me share an interesting experience that depicts how a natural, sincere and a systematic approach to a project saved me from the wrath of my superior officer when I was about to be framed.

The all-important project of construction of Break water at Andrott Island was sanctioned for implementation a few months before I took over the charge of EE at Kavaratti. I was to start the project as EE is responsible for effective execution of the project. I was going through the files to check what the design to be adopted for the same was. I could find none. So, when I talked to my boss on telephone that evening, I was mentioning that I could not find the approved design for the construction in the files and requested him to check the availability of the same in his office and send the same to me if available.

In the next two days, I got a long memo from my superior referencing the telephonic conversation and blaming me for being irresponsible by asking him the design. Further it read that it was my responsibility as EE to formulate the design and submit to him for approval. He had also added that already sufficient time had been lost and no more delay could be acceptable to him and if I did not submit the design details within ten days, disciplinary action would be initiated against me for non-compliance of instructions of superior officer, lethargic attitude toward work etc.

Shaken by this attitude, I learnt through other colleagues that he was one of those people who get work by threatening juniors. I was told that it was his habit to keep his subordinates tensed so that he could control them by fear tactics.

I immediately went on an exploratory tour to Andrott Island to get the designs completed. I halted there for a week, studied the site conditions and prepared a very comprehensive plan which included design, sequence of construction with drawings of each sequence, action plan with time frame and included the same in my Inspection Report of Andrott works and sent copies of the same to all concerned including my bosses. This time I made it a point that my boss's copy was sent by "Registered Post with Acknowledgement Due" to ensure that I have sufficient proof of having dispatched the details in time. The papers were received by the technical wing of his office; they processed it on file and put up for approval of design to my boss.

As usual, without going into details of all files, when he read the Assistant Engineer's noting that "Checked and found correct. Put up for approval please", he approved it which was intimated to me by his technical wing through official channels of communication.

That was sufficient for me. I started going ahead with the project activities as per my design and as per the action plan. Soon, the progress at the site was well appreciated by the local citizens of the island and it had also reached the attention of the Honorable Administrator of Lakshadweep Administration whom I had met on my arrival at Kavaratti on a courtesy call. During our conversations, I explained what I had been planning for this major project and shared with him my plans. He was highly impressed and wished me all the best and also assured all assistance from his administration.

In the next few days, my boss met the Administrator in Cochin to brief and update on our project progress. I found out later that my boss started bad mouthing me for my inactiveness and was trying to assure the Administrator that he would organize things shortly. But Administrator, who had just met me a few days earlier, snapped back stating that he had a detailed discussion with me and whatever I had made was the best suitable solution to the project and that my boss should extend all cooperation to me to execute as per plans.

One can easily imagine what was going to fall on me shortly. My boss being the last person who could tolerate any body appreciating his subordinates, sent me a long memo the very next day stating: I had not complied his instructions and submitted the design details of break water project within the specified date. This proved my inefficiency due to which the project was suffering. It further mentioned that the locals of the island were complaining against department to administration for which I was responsible. I was disallowed from contacting the Administrator without the permission of my boss. I should have apprised him of the status every day on telephone and that was treated as a serious lapse on my part.

As soon as I came to know that my boss met the Administrator, I sensed that something like this was about to happen as, by then, I had understood the psychology of my boss. I had prepared myself mentally to face it. So, I drafted a long and detailed reply giving all references of my sending him the details asked for by him, the date, the receipt no. of Registered post (enclosed a copy of the same), the letter of approval received from his technical wing conveying his approval etc. I also explained in my reply that I had spent sufficient time in Andrott Island during previous fortnight and interacted with local population who were very happy on the actions initiated there and in my presence, they had communicated their satisfaction of our progress to Honorable Administrator through telegram, which could be cross checked by my boss if he so wishes. I also mentioned in my reply regarding meeting with the Administrator, for which I replied that I would ensure that I would seek permission if I wanted to meet Administrator for any of our needs. Bu in case the Administrator called me to his office I would meet him with short notice, in the interest of our organization. I might not be able to contact my boss for his permission due to urgency, gravity of the situation and paucity of time.

The reply along with its enclosure ran up to seven or eight pages which contained all references very specifically. My reply was sent the very next day, again by "Registered post with Acknowledgment Due."

After two days, I called him on telephone and talked to him as if nothing had happened between us and briefed him of the progress of various projects including break water in a very cool voice contrary to his expectations. I made it a point to call him every alternate day and brief him. In case of any directions issued by him through telephone, I made it a point to put it in writing in the form of a letter, "As per your directions on telephone during our discussions on such date and time, the following points are noted for urgent action and compliance report shall be sent to you periodically without fail." This made him fall on his back foot and soon informed me told me that I was using the phone excessively and the phone bill was getting heavy and that I should be judicious with the official telephone. This gave me some reprieve from the every alternate day telephone calls.

Soon, my boss's approach changed. I used to get random telephone calls asking for work related details. When I used to explain the position, he started yelling stating that what I shared with him as incorrect and I should be more careful and attentive.

Confident that I was right, I started thinking how this can happen. I called my two Assistant Engineers (AE) to my office and casually enquired if my boss reached out to them over phone inquiring about the same details. Both of them confirmed that he called them that morning and was asking questions. Then, I understood the new technique being chosen by my boss to corner me. I immediately instructed my AEs that whenever my boss called them, we should get together as a team to discuss the conversation. This came into implementation with immediate effect. Within a week's time, my boss got frustrated of our internal arrangements and even asked me straight away if that was my arrangement. I honestly acknowledged affirmatively and thanked him profoundly stating but for him I would not have learnt these tricks to navigate the system. He laughed heartily.

Such experiences opened up my mind and thought process and make one more agile and competitive in work. This technique learnt from him was helpful to me when I was at the helm of affairs. As the head of the organization which spread over a vast area and scattered, I had to invariably be able to know what exactly was happening around in various projects and locations for which I had to adopt his techniques by contacting my confidante at various locations, cross check the information which was very essential for effective management. I always remembered him during those periods with grateful remembrances of old episodes.

3.5.3 Learning Office Politics from Scratch:

"Keep your thoughts positive because your thoughts become your words."

-L.S.S

Another great escape story now!

That was the time when Government of India had sanctioned the project of Construction of a runway at Agatti Island and was to be executed by a central government organization through eligible contractors. This project demanded transportation of all heavy machinery, construction materials including sand, quarry products, bitumen etc. from Indian mainland to Agatti Island. Being a underdeveloped island, there was no berthing facility for bigger ships carrying these items. This underlined the immediate need of floating barges (towable heavy floating hollow iron boxes fabricated with steel beams, angles and plates welded securely that can carry heavy loads of up to 100, 500 or even 1000 tons) on which they unload materials in the anchorage from mother vessel and tow it to the shore for unloading. We were the only organization with such barges.

Along with my boss, the General Manager (GM) of Airports Authority of India (AAI) also came to Kavaratti, Lakshadweep islands to kick start their project. When I went to see my boss at the Government Dak Bungalow, where they both were staying, the GM of AAI was introduced to me. Once he came to know that I was executive in charge, he requested me to spare the barges for their project of Agatti. I told him that my boss, who was staying in the same guest house would be the person to assign such resources and that he could directly talk to him. Upon his concurrence, I would gladly comply.

That evening, we three met there in the Dak Bungalow and he placed his request to my boss. He was overtly glad that he was being asked for help and he immediately directed me, "Mr. Sekar, it is our duty to help them. Please spare barges without affecting our project activity." The GM thanked him and we had dinner together there.

Fast forward to the end of the airport project. The project was completed by using our barges and by May 1988 or so, they were winding up from Agatti island.

These islands of Lakshadweep lie on the western side of South India, in the Arabian Sea and hence the effects of South West Monsoon are always severe. This monsoon starts by 15th May every year officially and closes by 15th September and all ports of Lakshadweep islands are officially declared closed by Lakshadweep Administration and there cannot be any shipping activity for non-sea going vessels. All smaller floating crafts owned by government departments as well as private parties would be hauled up on to the beach for annual maintenance as these cannot withstand the fury of sea. As per this practice, we too were planning to haul up all our barges before 15th May for maintenance.

AAI came running to me and requested to spare the barges at Agatti island till 20th May as by that time they would clear all their equipment or otherwise they would be stuck up for next six months without any use. I explained my plight and suggested to them to contact my boss at Calicut and if he agreed to it, then I could assist them. Promptly, the next day, my boss called me on telephone and said," AAI requested for barges till 20th May. Why can't we postpone our hauling up till then? "I said, "Sure, Sir, if you say so."

So, we spared two barges to them and they were using it for shifting back all their materials. One fateful day, the 17th, May, 1998, the weather had turned extremely worse and both barges which were alongside the ship in anchorage, started pitching and rolling like monsters and became uncontrollable. The captain of the ship snapped the ropes of the barges to save his ship from these barges as they started banging against the ship. Both barges started drifting and disappeared. Next morning, the news from Agatti reached me and my boss the same time. The next moment my telephone started ringing like a death bell. I knew what was in the offing. Yes. It was my boss. He started shouting in panic at the height of his voice. "What the hell is going on at Agatti? How were the barges given to AAI? Why have you not hauled them up on shore by 15th, May? If the barges are not brought back safely without damage, disciplinary action shall be initiated against you and the loss will be recovered from you. I am sending you a memo. You explain." He never allowed me to speak and I knew that very well.

My Assistant Engineer (Civil) and I set sail for Agatti the next day by a passenger ship which left from Kavaratti to Agatti and started our rescue operations by contacting all nearby ports through port communication system available at Agatti to enquire if they could cite any barge near their port. Finally, we could locate one barge got stuck up on the eastern shore of Agatti island and the other one in the shallow patches of Bangaram island which is near Agatti. Our next mission was to inspect them and to program for retrieval of both. Going to Bangaram at that rough sea conditions was ruled out and hence we concentrated on the one at the eastern shore of Agatti. We went and inspected it. It was half covered with sand which had accumulated over it due to change of tides and satisfied to note that there was not much of damage. With slight repairs, the barge could be operated like new. But the main problem was to retrieve the barge from its present position. We could understand that the pulling operation should be done only after clearing the sand covered over the barge and the whole operation of pulling it into water should be done between one high tide and corresponding low tide and at any cost shall not extend beyond that as raising tide would deposit more sand over the barge which makes operation most difficult and the whole exercise may prove futile.

So, we organized ourselves as follows with innovative technology utilizing locally available tools and plant and materials.

-Collected spades and like materials in large numbers for clearing the sand in one go
-Organized sufficient manpower
-Collected two numbers of hand winches (20 ton capacity each) from some other department. (Nothing more was available with anybody)
-10 Ton capacity cast iron pulleys
- Available quantity of D Shackles (Cast iron in the form of locking shackle in the form of D with screwing arrangement for locking)
-Steel wire ropes used for pile driving earlier during construction of jetty.
-Prepared a few wire slings out of the cut pieces of the wire rope
-Dug a big pit on the shore to burry a strong anchor to retain the hand winches in position when the barge was to be pulled.
-Fixed the anchorage properly and securely and located the hand winches in vantage positions so that the barge could be pulled towards the sea when ordered.

The self weight of the barge was at least 70 Tons whereas we had only 2 Nos. of 20 Ton hand winches. So we had to organize coupling of pulleys to gain mechanical advantage so that with available less power, we could finish the operation successfully.

Early morning of the next day, when the high tide started receding, we too started removing the deposited sand putting all our might, arranged to tie the slings around the barge, secured it with D Shackles, connected the system with wire ropes running from the hand winches through the sets of pulleys to the slings around the barge. By low tide, we removed all deposits over the barge and started the pulling operation. In spite of all our efforts, the barge was not moving because of the friction underneath. So as a quick thought, we had collected a few trunks of felled coconut trees which were available nearby and tried to insert underneath the barge. It started working. The barge started moving downwards and by the time the tide was high, the barge started floating to the celebration of all local population who had gathered there to witness the whole operation. They were so helpful and appreciative that their local leader had come forward to offer tea to all who were involved in the operation. It was a happy occasion to all of us.

The second barge at Bangaram shore was inspected by getting reaching there with local fishermen when the sea was not so rough. We found that there was no damage to it. It was securely tied to the nearby tree to retrieve later when the sea turned calm.

We returned back to Kavaratti the next week, where another strong Memo was waiting for me from my boss as he had already promised (!). The main concern areas were:

- No proper permissions to release the barges to AAI
- Lack of timely action on my part to haul up the barges on or before 15th May.
- Explanation as to why disciplinary action should not be initiated against me for the unresponsiveness on my part

I thought for a while. Then called my Assistant Engineer who by then developed a great amount of regard and respect towards me after our Agatti operations and requested him to go to Kavaratti Dak Bungalow. There I requested him to search the old register of guests who stayed during the particular period when my boss and GM of AAI were staying there and if possible try to retrieve a copy of the registration page with both my boss's name and AAI GM with dates of check in and check out. Within an hour, he promptly brought me a copy of the page from the register. That established the clear dates on which the first episode of meeting between my boss and GM of AAI.

This alone was not sufficient to build an air tight case. I needed proof on paper that I released the barges as per my boss's instructions.

So, I thought of a plan. I wrote a personal letter to the GM of AAI (who knows me personally as we had met on a few more occasions) congratulating him for successfully completing the project at Agatti. I added that my boss was also remembering him last week and probably as a courtesy, GM may drop a letter of thanks if he felt so, as it was due to the directions issued by my boss on the request of GM during our meeting Dak Bungalow, Kavaratti, that the project could be completed successfully.

In two days I got a call from GM of AAI who profoundly thanked me for reminding him correctly and asked for the postal address of my boss for sending him a letter of thanks. I happily gave him his address and also cautioned him not to thank me in his letter as generally it would shift the focus of the letter. But tactfully, I made a casual remark that perhaps he could mark a copy to me when he sent the letter of thanks to my boss if GM felt like. Within a week, I got a copy of his letter addressed to my boss by post. The original letter too reached my boss but he did not say a thing to me.

Now that I got a proof on paper that the barges were issued to AAI on the directions of my boss, I prepared a comprehensive reply enclosing a copy of the letter received from GM of AAI giving all possible references to the extent. To add to that, both barges were retrieved safely, there was really no case against me.

After a few days, I called my boss on the phone and enquired if he had received my reply to his memo. He laughed outwardly but spoke nervously, "Sekar, yes, I did get your letter but I could not read it as I have been busy. Why don't you come to Calicut for discussing and closing the issue?"

Next week, I went to his office at Calicut. Before meeting him, I was talking to our people in the office when one officer confided in me "Sir, so far nobody stood back to our boss, and he thought it was always his prerogative. But now you have given him cause for worry. Please be watchful of him sir". I really appreciated his courage in telling me that.

Then I met my boss in his office chambers. He was most cordial and invited me enthusiastically. Then he pulled out the letter sent by me from his table drawer, the cover still un-opened. Before I asked him, he told me that he was nervous to open and read that letter and so he did not open it. I said it contained only facts and nothing to worry. He replied, "I know that.

That is why I was worried. The earlier executives before you were afraid of memos and they were always scared of me. They were not good in sending replies. But unfortunately you are good in replying and that too, with all references, which frankly bothers me."

I was glad at his honest admission in private and shared with him that I had to do it to save myself from such situations from him. The responses were only necessitated when I was being issued memos. I also took the opportunity to tell him that it was easier for me to respond to the memos as they were superfluous and lacked specific items with references. He laughed with me and agreed that he would not issue any more memos as it eventually turns back at him. He decided to sort out issues by discussing with me personally.

He hated boomerangs!

One important theory of management which I learnt from him is that whenever you suspect that someone is likely to create trouble for you, he or she may do it only when his/her mind is free. Never allow them to be free. While keeping all of your actions on record and as per procedures, try to keep problem mongers busy.

An empty mind is a devil's advocate.

Concentrating on their own issues and concern areas will keep them off your back. This is an unwritten theory in management. But I would definitely like to draw a comparison to one of the great epics of India, the Mahabharata and the life of Karna in particular.

In the great epic of Mahabharata, Karna who was on the side of Kauravas was undoubtedly the best warrior and the best archer among the contemporary which had made the Pandavas and Lord Krishna nervous and weary. He was perhaps considered even better than Arjun in archery. Lord Krishna decided that he would somehow control Karna's powers. He then sends Karna's original mother Kunti to him (who was also the mother of Pandavas) to identify herself as mother of Karna for the first time and ask for a boon from him that he would not kill any of the Pandava brothers in the Mahabharata war and secondly, he would not apply "Nagastra" for a second time on Arjun.-Willful Injustice 1&2

Earlier, Lord Indra goes in disguise of an old sage to Karna and gets his Kavach and Kundal which were born with him as protection from Sun God.-Willful injustice 3

Due to curse of his Guru Parasuram, Karna forgot the mantra of Brahmmastra when it was needed for him.- Injustice 4

Lastly, when Karna was badly injured and lying down in the Kurukshetra battle field, Krishna goes to him in disguise as a Brahmin and begs for all his virtues of charity and de-protect him from the Goddess of Dharma. Injustice 5.

Imagine a situation that all these five injustices were not done to him. The story of Mahabharata and the end of Kurukshetra war would have been totally different.

What one can infer from this is "Disarm a person in as many ways as possible if you want to emerge victorious". My earlier boss seems to have followed this very philosophy. I, in turn, learnt this from him to keep trouble creators at bay and protect my organization throughout my tenure.

I am really thankful to him for exposing me to his management philosophy. These instances not only taught me to be watchful of a new and dirty way of management that existed out there and also helped boost my self-confidence by handling such situations of pressure.

3.5.4 Salvaging of Sunken Barge Annapoorna:

"Don't try to teach anything to anybody. Try to make them think."

After Lakshadweep, I was posted to Great Nicobar island on transfer, a difficult place to live because of its extreme remoteness. Great Nicobar is the Southernmost island in the Andaman Nicobar group of islands. As soon as I assumed charge as an Executive Engineer (EE) at Camp Bell Bay, the very first task awaiting me was salvaging a Hopper Barge(barge – a towable heavy floating hollow iron boxes fabricated with steel beams, angles and plates welded securely that can carry heavy loads of up to 100, 500 or even 1000 tons) which had sunken near an island called Kondul, just North of Great Island.

The construction of a jetty at Kondul Island was on progress at that point of time for which all construction materials like steel rods, cement etc. were to be transported from Camp bell Bay harbor by loading them on barges and tow the barges up to Kondul. Depending on the sea conditions on the way, the travel time could be six to eight hours one way. During one such transportation, the sea had become extremely rough and the barge had snapped off from the tug, drifted into St. George Channel between Great Nicobar and Kondul.

The channel being very narrow experiences heavy water currents flowing back and forth at every change of tide. The water depth at the location where the barge sank was about 8 meters. It was imperative that the barge was to be salvaged and commissioned immediately as almost all the projects of our organization were dependent on it. With very little flexibility in schedules, we had to plan our project activities within the available resources.

My immediate step was to take an inventory of available resources in terms of materials, machinery, equipment, manpower and my team of technicians and Junior Engineers. I did manage to collect the following items from our central stores, other departments, etc.

Dumb Barge

- One more barge in floating condition for use as working platform
- Diesel Engine driven air compressor-1 No. with hoses
- Diver's helmet – 3 Nos.

- Diesel Welding Generator – 1 No. and sufficient quantity of welding rods.
- Hand winches 10 ton capacity – 2 Nos.
- Dewatering pumps 2Hp 2 nos.
- Steel stranded wire ropes – about 1000 meters
- Manila tow ropes – 2 big bundles and ordinary life ropes
- Triple sheave cast iron pulleys 4 sets
- D Shackles of various available sizes
- Crates of mineral water to drink
- Paper plates, paper napkins for food
- Small quantity of cooking items and groceries
- Two drums of diesel
- Other miscellaneous items like tool boxes, cutting tools etc.

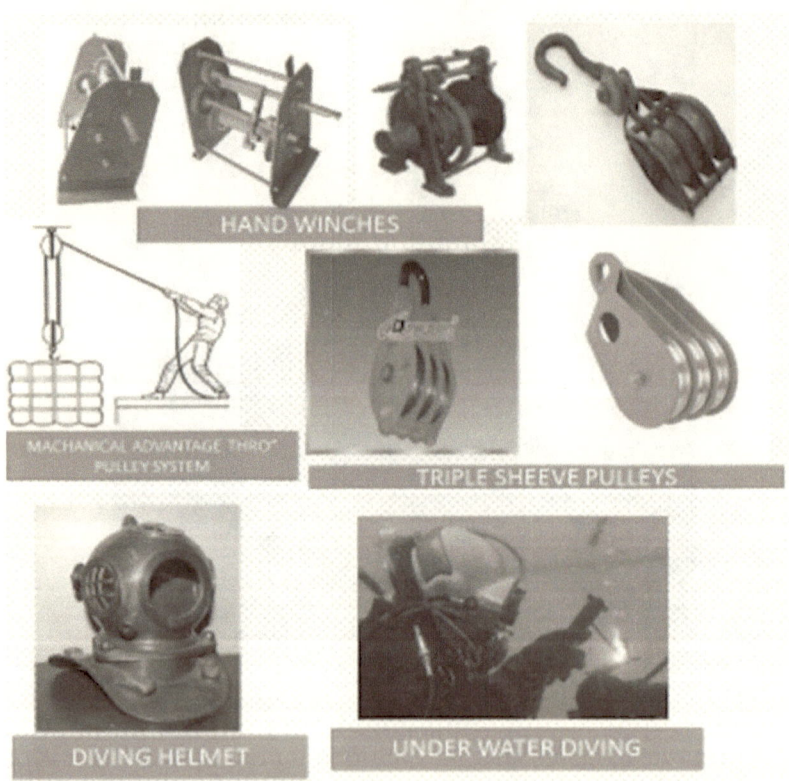

HAND WINCHES

MACHANICAL ADVANTAGE THRO' PULLEY SYSTEM

TRIPLE SHEEVE PULLEYS

DIVING HELMET

UNDER WATER DIVING

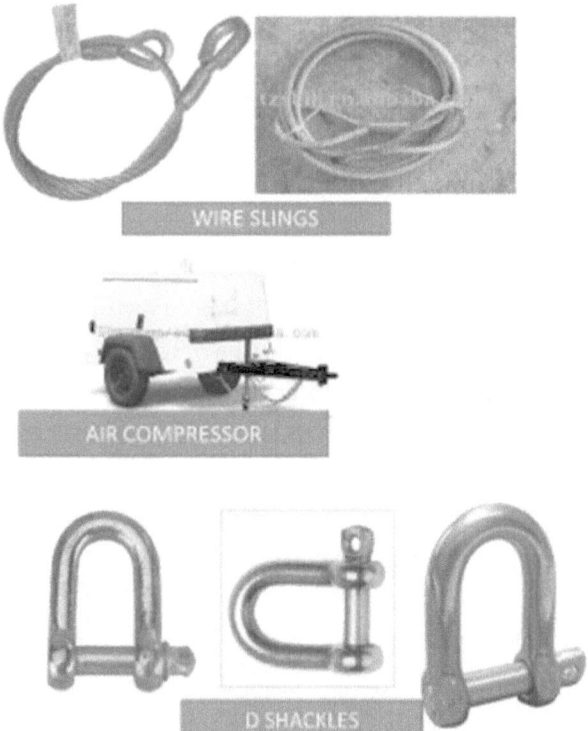

I then shifted focus on selecting my team members who were willing to brave the adverse conditions at Kondul. With no electricity, no proper food, untimely work hours which depended on the availability of low tides which could be night, early morning and sometimes during day too and infested with lots of malarial mosquitoes after dark, this was one of the most challenging work destinations for my organization and my team members.

I needed at least two good operators for the machinery, two divers who could dive down into water alternatively to inspect, fasten, etc. below water at that depth of 8 meters, braving the cold and the dreadful current. I also needed supervisory staff who could quickly understand the situation, my directions and act smartly without wasting any time which was of the essence.

Finally, the people who stood courageously with me to put themselves through the test were Mr.T.N.Krishnamoorthy, Assistant Engineer(Civil), Mr. Sivakumar, JE(Mechanical), Mr. Kasinathan, JE(Civil), Mr. Chellam, Senior Foreman (Mechanical), Mr. Manickavasagam, Senior Foreman

(Civil), M/s Alagu and Gandhi, both under water divers. Along with us, we had aman power of about 10 people for this operation.

After collecting all the items, we all set sail for Kondul on "**Operation Annapoorna**" after the name of the sunken barge whose name was Annapoorna, with a tow of another barge similar to Annapoorna (our working platform) behind us.

Our team reached the location where the barge was lying underneath the water at the St. George Channel at 1200 hours, with bright sun on top of us, calm sea and mild breeze. When we were trying to locate our target, it was so clearly visible in the prevailing clear weather.

Not wanting to spend an unproductive minute, I wanted one of our divers to inspect of the extent of damage to the barge. Mr. Gandhi was so enthusiastic. He changed to his diving costume and was about to dive into cold water. At that last second, I stopped him and pointed out that he had not fastened his life line rope around his waist failing which he was not going to be permitted to be a part of this operation. With all hesitancy, he tied the rope in his waist with the other end of the rope secured safely.

On the word go, Mr. Gandhi dove into the water towards the sunken barge. The very next moment, there was a huge screams from everyone aboard our tug. Mr. Gandhi was nowhere near Annapoorna but was carried away by the underwater current at least by a distance of 10 meters but was held tightly by our staff who clung on to the other end of the life rope. Mr. Gandhi was pulled up over the tug immediately. On reaching the tug, he looked at me with folded hands expressing his gratitude. This incident at the beginning of the operation itself made every one serious about the task at hand and was self-explanatory around the necessity of self-discipline and to stick to plans. Activities were allocated to various small teams like:

-Arranging and fixing of two hand winches at the centre edges of two sides opposite to each other in our working platform barge
-Spinning of large slings from the steel wire rope pieces which would be used to tie around Annapoorna under water.
-Greasing and oiling of the pulleys and making serial connections of pulleys with insertion of wire ropes suitably to gain mechanical advantage. (This was very important for the reason that the

self-weight of Annapoorna was about 60 tons (60000 Kg.) and it had water in all its compartments)

-Providing safety arrangements for our team especially due to the space constraint on the working barge as all equipment too were to be mounted on the barge

-Checking the welding generator, air compressor, dewatering pumps, add fuel in them and make them ready for operation

It took about two full days for preparing all the above. One thing I would like to mention at this stage is that there was absolute harmony and coordination on these arrangements. Every one of us was in our half trousers to acclimatize ourselves in these physically demanding work conditions. There was absolutely no feeling of senior, junior, officer, labour etc. I didn't feel like the leader of the operation anymore. I felt like one among them. Every body's mind was laser focused on the challenge at hand. I was pulling the wire ropes with labour force, helped in making slings, even served tea to my team on board barge and everyone was feeling great to be a part of this salvage operation.

As it was already late and the sun was setting, we called it a day and decided to go in full swing from the next morning. We reached our site office which was a small old house with two rooms, a kitchen and one toilet. There was no power and so we had only kerosene lamps. After a wash, we rested for a while. Then food preparation was started by our staff for all of us. Meanwhile, I saw one of our staff was moving suspiciously. I called him and enquired and found that he had collected a few bottles of 'toddy', which is an unregulated alcoholic drink made out of extracts from coconut trees. This cools down the body and has negligible intoxication. I immediately told him that it was not fair on him to have it alone and said that all of us would have it together. He was so surprised and shocked and shared his sentiments with others saying he was yet to see an officer who was so thorough with work at the same time never had that ego of an officer and one who treats team members like friends. I was very happy to hear that comment and felt very good thinking to myself that it was in my nature and did not require any special effort to come across that way.

Next morning, we got up early, had some tea, adorned our half trousers and proceeded to our barge. Anyone who knew lesser would have thought a bunch of friends were going for a picnic on the beach. Little would they have imagined the uphill battle we are all fighting!

Our first task was to send the diver down to check the dents and holes inflicted on Annapoorna. The air compressor was started, the diving helmet was connected to the air compressor with hose and tested. The diver was initially told to wear the helmet and go just below water (of course with life rope) and check if it was comfortable for him to breath. Both divers took trial and slowly their diving depth was increased and finally they could reach Annapoorna, walk on it and could concentrate on its problems.

While this was going, the wire ropes were fastened to the hand winches at both sides of the barge and connecting the pulley blocks was on.

After initial inspection of the chambers of Annapoorna, it was learnt that two out of six compartments had developed holes and the man-holes were not fixed properly.

Our next attempt was to plug the holes in those two compartments by applying dry cement and then try to dewater from those compartments using the suction pumps. The idea was, if this could be done effectively, the compartments may start lifting once the water was pumped out (as the emptied space will have only air) and this would help in towing the same when it was above sea bed level, by suspending it from the barge above with wire slings prepared for the purpose.

When we tried applying dry cement to plug the holes, it was not successful as the sea water was not allowing for setting of cement.

So, we concentrated on the compartments which were not damaged. We fixed the suction hose extension to fit with the man-hole tightly and started our dewatering pumps which were mounted on the floating barge. Soon we could see the results. Annapoorna started to be buoyant and we could get some clearance between Annapoorna and the sea bed as it started lifting.

Our next plan came into force. We channeled the slings around Annapoorna, fastened it to the wire ropes through the pulley system which was powered by the hand winches.

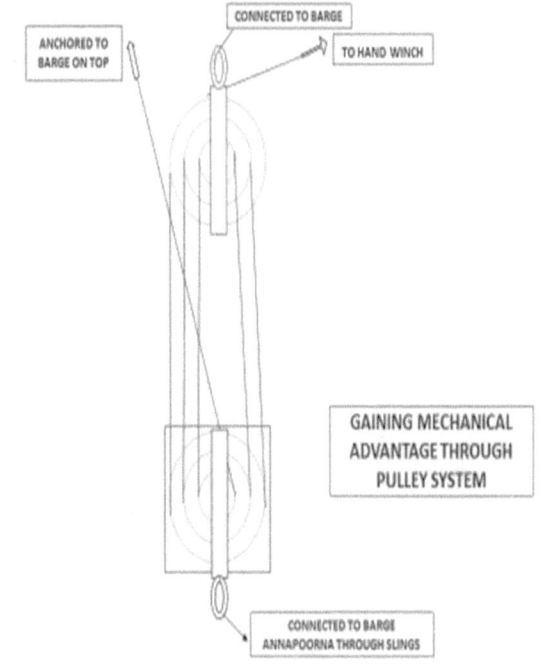

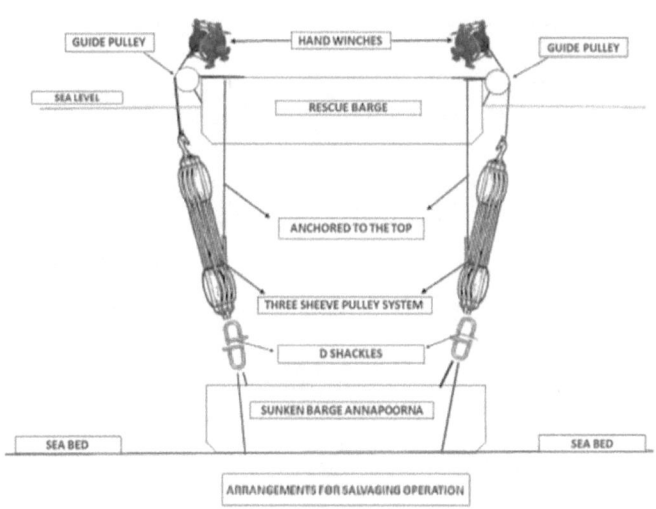

Once the connections were confirmed, the hand winches were slowly operated from both sides carefully with perfect coordination.

Imagine manual labor being used to lift a 500 ton beast!

Initially, there were some crackling noises when the ropes were getting tightened but stabilized shortly. Soon Annapoorna started lifting herself from the sea bed like a baby in the cradle.

Once Annapoorna had a clearance of at least a meter from sea bed, we held it secure and tight. We passed two towing ropes, one from Annapoorna and the other from our barge to the tug which was waiting nearby.

When this was done, it was about 1600 hours that day and the climate was moderate. Our mission was to tow ourselves around the corner and reach Eastern side beach area of Kondul. The tug started towing us. Initially we were thrilled to see that we were moving as a package, - the tug in our front, Annapoorna hanging below us and we on the barge.

As we approached the corner where we had to turn left to reach the eastern coast of Kondul, the wind suddenly started to blow very heavily against our target and the waves were rising severely. Our barge with Annapoorna was tossed up and down heavily and the ropes were making heavy and dangerous crackling noise when it rubbed against the surface of the steel barge. All our team on the barge was so tensed and worried thinking if we would reach shore safely and see the light of the next day.

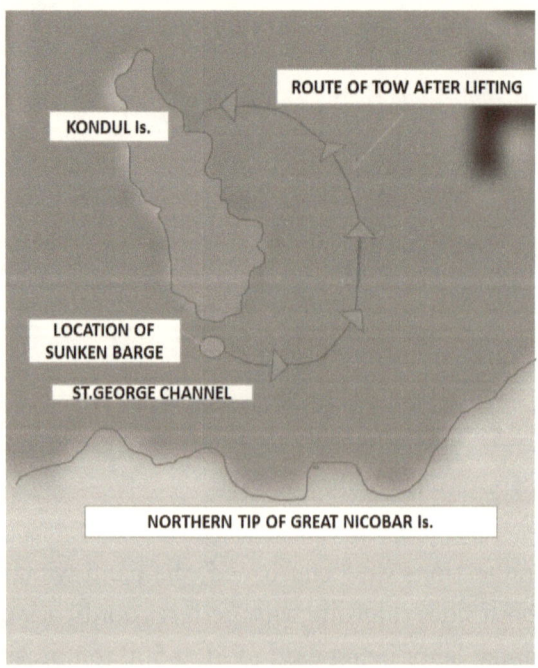

The situation was beyond our control and the only thing we could do then was to pray to heart's content. I told my team we were about to have a memorable experience that we would remember forever in our life. I asked those who believe in God to pray instead of panicking.

By then it had become dark and the tug was pulling with all its might without any positive motion towards front due to heavy wind against us. After half an hour of this life struggle, we realized that we were moving towards the East coast of Kondul. After we turned through that corner there was no further problem and we could safely reach our destined location.

After securing the barges safely, we went back to our site office and had a grand party, of course with toddy, chapatti and dhal. What else could we have? Remember, this is one the remotest island with very little facilities.

Next morning, we undertook some temporary repairs to Annapoorna, brought it up to floating condition and towed it to Camp Bell Bay for attending to major repairs. The whole team was feeling great as it was an achievement for them which were to be cherished throughout their life. Needless to say, I was very proud of what we had achieved with meager resources!

After this, Mr.T.N. Krishnamurthy, AE stood by me was with me in yet another salvaging operation of a similar barge at Katchal islands after a few months which also was a huge success.

Any sincere effort with due perseverance and dedication coupled with timely action will yield success without fail. All the more important is that the team leader should be capable of gaining the confidence of the team members and should lead from the front during times of crisis.

When I rewind my memories on this episode, I feel elated that I could live up to the expectations of my team and in spite of not having any formal management education, was applying the principles of effective management by instinct. Every manager can develop that instinct by observation, reading, being aware of oneself coupled with determination to perform and deliver.

3.5.5 Administrative Debacle:

> *"Don't be in such a hurry to condemn a person because he doesn't do what you do or think as you think. There was a time when you didn't know what you know today."*
>
> *–Malcolm X*

I would like to share a few more interesting incidents from my career which throws light on how things fall into disarray when the manager does not take timely decisions.

This dates back to the time, I was asked to undertake the additional charge of Administrative Officer as the post was vacant. We had about 20 to 25 administrative staff working in the office. I received a request for a good ink pen from the staff as there was no stock of pens in office. So, I had submitted a note on file for approval of my boss to procure 25 micro-tip pens, one for each staff in the administrative office. I had also included the quotations obtained from various suppliers, comparative statement, recommendations etc. as I knew that there would be lots of queries and if full details were not provided, a lot of questions would be raised on the file.

The file came back with a query inquiring about the eligibility as per government manual for providing such pens to staff. I never ever imagined that such a query would be raised. Anyway, now that it was raised, it had to be cleared. So, I had marked the file to Office Superintendent (OS) with a request to verify with manual and clarify. The file came back from OS with the remarks, "Verified with codes and manuals. There is no mention of eligibility on such issues. The competent authority may take decisions."

Then I added my remarks as, "As the Head of Office is the competent authority, he may kindly approve." and sent the file to my boss.

Quick came back the file with remarks, "Please indicate provisions kept for purchase of pens in any estimate sanctioned."

I wrote back, "Purchase of these items fall under Office contingencies for which separate fund provision is there. Details of fund availability, Flagged as "A".'

"Seen. As this pen seems to be costly, one pen may be purchased for use of Administrative Officer."

Me: "The request was to provide pens to all the 25 administrative staff and not just for Administrative Officer. Hence, if approval is accorded for purchase of only one pen, then it is worth to be used by boss only as it is so costly and precious. Submitted for approval only."

The file never returned. When, it came to my hands after a month or so, those pages of note sheets pertaining to purchase of pens were not at all seen in the file!

This experience shows how a government office functions and how a boss should be proactive in taking decisions and avoid unnecessary wastage of time. I would narrate another incident with the same boss.

3.5.6 Completing Construction of Slipway, Mayabunder- Constraints overcome:

> *"Keep your words positive because your words become your behavior. Keep your behavior positive because your behavior becomes your habit. Keep your habits positive because your habits become your values. Keep your values positive because your values become your destiny."*
>
> *– M.K.Gandhi*

A project of Construction of a slipway was on at Mayabunder, in Northern Group of Islands in Andaman and Nicobar Islands. A slipway is a type of ship repair facility in the form of a sloped rail track sloping from the sea towards land. The ships requiring repairs are seated on a trolley at the sea end and hauled upwards with the help of power winches on the land for repairs in dry condition. The officer working on the project earlier completed the easier part of construction which was above water level but could not achieve the construction below water level as there was seepage of water all the time. It had become a challenge and the work was pending for more than a year. I was called upon for completing this project on an urgent basis.

Slipway at Port Blair

I planned to execute the remaining project work in parts by compartmentalizing smaller portions in the difficult area where water gushing was always a problem due to proximity of the sea. This way the seepage water could be kept under control by dewatering the seepage water continuously by engaging multiple number of dewatering pumps so that concreting could be done using quick setting cement. For making compartments, I needed about 3000 empty cement bags which would be filled with sand and stacked to make internal bunds for compartments. As this quantity was not available with the department, I had called for short quotations for this item. We had received 5 quotations and only two of them complied with the time schedule of supply. The order was placed and the items as required were procured in 4 days. These were transported in the department truck going to Mayabunder with some other construction materials of the department the very next day.

Meanwhile, my boss called me to his chamber and said that he had received a complaint on me that I had favored an agency by giving higher price to empty gunny bags on the said purchase. I told him that my files regarding the procurement of this are open for scrutiny and immediately called for the complete file and handed over the same to him. After going through the file, he observed that there was another agency who had offered a rate at 50 paisa less for each gunny bag but I had not considered his rate.

I explained to him that he did not comply with timeline of supply as per quotation notice which clearly stated that those who cannot supply the required quantity within 7 days from the date of acceptance of the quoted rates need not participate in quoting and such quotations which do not comply with this condition shall be summarily rejected. I had also indicated that this agency had mentioned the delivery period as 3 months which is not acceptable for us and hence his quotation was rejected as his quote was not in compliance to conditions mentioned therein.

I was very clear in my action. But, as usual, he was not in a position to reject my point and at the same time too scared of the complaint also. So he asked me to place orders on the agency who quoted lowest by cancelling the order issued.

I explained to him that this agency had already supplied the materials and they were already transported to Mayabunder. And we were ready to start the work in 2 or 3 days. He was not seen convinced. So I told him, "Ok Sir. I will return the bags to the agency who had supplied. I will cancel the order issued to him. But as per procedure, please allow me to cancel this round of quotations. Let us call for fresh quotations removing the time limit for supply which may lead to still lower cost too."

He shot back stating that the work would be inordinately delayed.

I agreed and left his chamber saying that I would wait for his decision. Before leaving, I asked him whether I could arrange transportation of those bags from Mayabunder as our department truck was returning to Port Blair the next day.

He understood he was in a fix. After thinking for quite some time, he told me "No. Don't bring them back. Go ahead with the work. After all there is no mistake done by buying bags based on urgency at a marginally higher price when no one else could supply the quantity within the specified period."

I retorted agreeing, "Sir That is what I have been explaining to you for the past half an hour. Thank you Sir."

The best way to tackle such complaints in government departments is either reverting back with all air tight defensive proof on record (provided you have the support of your boss, the department and the controlling ministry which is most unlikely) or simply yield to the pressures. The decision making ability is directly proportional to the support you get from your superiors as well as your subordinates for which you have to prove yourself right logically first and in all practical aspects of your job subsequently.

After the initial debacle, we went ahead with the project every single day based on the timings of low tides. I engaged a large band of manual force, 13 dewatering pumps; additional generators as power back up and completed the project successfully while many people kept prophesying that we will fail in our efforts.

I always believed that hard work coupled with sincerity and a little bit of brain never fails. This experience gave me another team of excellent and enthusiastic team members. One team member I would like to mention is Mr. Abdul Rehiman, Junior Engineer(Mechanical) who was the key person running and maintaining 13 dewatering pumps without which the project would have never succeeded.

It is very important that the senior level managers should be capable of quick thinking and fast decision making failing which the whole team may get demoralized thereby putting the project in jeopardy.

3.5.7 Cute planning- Construction of RCC Bridge across Austin Strait:

"When it is 'RIGHT', you feel it. You don't question it or convince yourself. As biting into a lemon is undoubtedly sour, the feeling of something being "right" should be felt undoubtedly too."

-LSS

Another incredible achievement was the execution and completion of another project which is Construction of an R.C.C.Bridge across Austin Strait connecting Mayabunder and Diglipur in North Andaman. This project was a pet project of Andaman and Nicobar Administration. By connecting the said locations by a permanent bridge, the Andaman Trunk Road (ATR) would connect Port Blair in South Andaman to Diglipur, the Northern Most island in the North Andaman by a motorable road length of 333 Km. This prestigious project was handed over for execution by A & N Administration to ALHW.

After detailed studies and investigations, special construction techniques were planned and organized. The project site was about 7 km from Mayabunder and was of course, extremely remote such that it was not possible to get a cup of coffee or tea. It was decided to deploy barest minimum staff who were made with sterner stuff to handle this project from the site with all technical and logistical support extended from the Division office at Rangat which is at a distance of about 55 km.

My deployment at site for this project consisted of 1 Assistant Engineer (Civil), Junior Engineers (Civil) couple, (Yes! they were husband and wife), 1 Work Assistant, 1 Lower Division Clerk and a Watchman.

Austin Bridge at Mayabunder

That was all from the department. The contract vendor agency who was awarded with the work also followed the similar guidelines in their deployment of work force. They were staying at the temporary buildings at site and they knew that the earlier they finish the project, the sooner they get out of the remoteness!

Thus started the project activities at a very high pace while everyone had their apprehensions around my decision on deployment. Many of them openly lamented about how an important project of that magnitude could be executed by such lean staffing. I was extremely clear on my plans, decisions and my team's execution.

Finally, we were able to complete the project three months ahead of schedule to the appreciation of all without any cost overrun.

Clear cut thinking, planning, visualizing the project from a realistic perspective and above all team work were the main reasons of success.

3.5.8 Live Up To Other's Confidence in You:

"God changes caterpillars into butterflies, sand into pearls and coal into diamond using tome and pressure. He is working on you too."

–Rick Warren

That was the year 2001, when Port Blair City was experiencing acute water shortage and the Andaman and Nicobar Administration was running pillar to post to solve this problem. Several experts were called in to locate fresh water sources which had not been explored so far then. One such location identified was at Rutland Island which is uninhabited and located at a distance of about 2 km (across the sea) from Chidiatapu, a location on the Southernmost part of Port Blair.

In Rutland island, there are seven perennial nallahs (Rivulets) carrying potable water and flowing into the sea. If all the nallahs were integrated, there was ample water but it was to take sufficient time and money to be invested. I was invited to join a meeting conducted by the Chief Secretary,

Dr. Padmanabhan,IAS, who had a high opinion of our department's working and technical and managerial skills.

During the discussions, he was interested in knowing my views. I said, "Sir, Let us do what could be done on an emergency basis to meet our immediate requirements to maximum possible extent. I suggest that we tackle one nallah first; construct a temporary check dam, say about a meter high, lay flexible pipelines from the check dam and extend it into the sea by providing temporary supports to a point where low drafted ships and self-propelled barges can reach in anchorage, provide one diesel operated water pump to pump water from the check dam into barges or ships directly through the pipelines."

He liked the idea and asked if Andaman Public Works Department (APWD) could execute it on war footing basis. When the Engineers of APWD were still reluctant to commit, the Chief Secretary immediately intervened that this work would be done by ALHW and that I should take up the project on behalf of my organization. He asked me for an estimate of how many days it would take us to complete the mission. After detailed discussion, we settled on a 15 day time frame.

I deployed a dynamic team and completed the work as planned in 10 days and went to the Chief Secretary with the photographs of the completed constructions. Seeing the big jets of water from the hose pipe end into the sea, he was extremely thrilled to realize that the project turned into a reality, that too ahead of schedule! He announced that he would come to commission it the very next day with other secretaries.

I was thrilled at his enthusiasm to come to the location as it was approachable only by country boats with small out board engine. When I explained these aspects to him, he said, "Mr. Sekar, When you have done it for me, I will do it for you. I am coming tomorrow!" I was very much moved by his gesture.

He came, he saw and we conquered everybody's heart!

SECTION 4.0

EXECUTIVE LEVEL MANAGEMENT

4.0 Executive Level Management

"You cannot change the people around you. But you can change the people you choose to be around."

- LSS

It is extremely difficult for anyone to fit in at the level of Executive as it requires a lot of refinement, maturity, understanding, balance of mind etc. which are gained through experience from Novice level and Middle Management level. Having obtained most of the qualities required for Middle Level Management, it is quite obvious that the next aspiration is to become the best Manager in the Executive Level. An aspiring Manager in the Executive Level shall be a great Developer of the existing talents in him as well as around him.

He is capable of identifying individual talents around him, pick it up and apply at the right sphere.

He shall have a strong Focus on the talents, targets, route plan, when to do, what to do, how to do etc.

Though he has all the command in him because of his clear vision of the project, he is also deliberative as he believes that he can find a great idea from any one at any level.

He organizes himself very well by arranging things in a coordinated way so that it is easy for his team to go ahead smoothly.

He acts as a perfect catalyst that activates events.

He is highly responsible of the task in hand which he moves carefully and strategically.

Different important qualities of a successful Executive are discussed below.

4.1 Team Developer:

> *"Ego is a thing which makes you nonexistent if it is not under genuine control. You may decide if your ego is important or you achieve something by controlling it."*

> *-LSS*

When you manage a team, how well it performs often depends on how well you've trained and developed your people.

Once, I was walking along the road in our colony in the morning. I saw two workmen on the road side doing something. When I observed closely, I was shocked and upset. One person was digging a pit of about 1 foot by 1 foot by 1 foot at a distance of about 10 feet on the side of the road and the other person was, shockingly, filling those pits by putting the soil again into the pits. As I could not hold my anxiety, I went near them and asked what they were doing. One person replied that they are a team of three persons. His job was to dig pits at regular intervals along the road side, the second person was to plant a tree in the pit and the third person was tasked with filling the pits. As the second person was on vacation that day, the other two were doing their duty. Appalled at their version I inquired about their supervisor. They shared that he was out having breakfast.

This is a glaring example of a bad team work. None of the team members understood the collective objective of their team. The team leader, the supervisor in this case has failed miserably in his role of a manager and more importantly, a leader. I have used this example with my staff to caution them against the pitfalls of poor leadership. On the other hand consider this scenario.

Three stone-cutters were engaged in erecting a temple. As usual a Human resource development consultant asked them what they were doing. The response of the three workers to this innocent-looking question is illuminating.

'I am a poor man. I have to maintain my family. I am making a living here,' said the first stone-cutter with a dejected face.

'Well, I work because I want to show that I am the best stone-cutter in the country,' said the second one with a sense of pride.

'Oh, I want to build the most beautiful temple in the country,' said the third one with a visionary gleam.

Their jobs were identical but their perspectives were different. What Bhagwat Gita tells us is to develop the visionary perspective in the work we do. It tells us to develop a sense of larger vision in one's work for the common good.

Team members need ongoing training and development to help them become more effective, and take on progressively challenging activities. More than this, they need help learning new skills as the nature of their work – and that of your organization – changes.

Too often, companies limit training and development to new hires and to people moving into new roles. This is a mistake, because ongoing training helps people adjust to changing job requirements. It also creates a pool of qualified and available pool of people, who are ready to step into new roles as your organization needs them. This process will help the executive develop a more effective, efficient, productive, and motivated workforce. Done properly, this will ensure that the executive achieves his/her objectives and improves their competitive position.

More recently, organizations have come to understand that leadership can also be developed by strengthening the connection between, and alignment of, the efforts of individual leaders and the systems through which they influence organizational operations. This has led to a differentiation between 'leader development' and 'leadership development'.

The development of "high potentials" to effectively take over the current leadership when their time comes to exit their positions is known as succession planning. This type of leadership development usually requires the extensive transfer of an individual between departments. In many multinationals, it usually requires international transfer and experience to build a future leader. Succession planning requires a sharp focus on organization's future and vision, in order to align leadership development with the future the firm aspires to create. Thus successive leadership development is based not only on knowledge and history but also on a dream. For such a plan to be successful, a screening of future leadership should be based not only on "what we know and have" but also on "what we aspire to become". People involved in succession planning should be representative of current leadership representing the vision and HR executives having to translate it all into a program.

You're doing many of the right things to develop and train your people. Take it to the next level by making staff development a priority. Think about creative ways of sharing knowledge and inspiring your people to improve their skills on a daily basis. The more you show that you're committed to their long-term success, the more motivated, satisfied, and productive they will be. And that will make your job much easier.

Hold regular one-on-one meetings with staff to discuss and understand people's developmental needs. In these meetings, explore their current performance, and identify areas for improvement. From there, create a development plan to fill any skill gaps and prepare the team member to meet the challenges ahead. This is where it helps to have a competency framework for each person's role, and it's where it's worth conducting a training needs assessment to identify the training and development that each person needs.

Reaching the end of a job interview, a Human Resources Officer asks a young engineer fresh out of school about his salary expectations.

The engineer replies, "In the region of $125,000 a year, depending on the benefits package."

The interviewer inquires, "Well, what would you say to a package of five weeks' vacation, 14 paid holidays, full medical and dental, company matching retirement fund to 50% of salary, and a company car leased every two years, say, a red Corvette?"

The engineer sits up straight and says, "Wow! Are you kidding?" The interviewer replies, "Yeah, but you started it."

Successful teams and organizations typically put a lot of effort into developing future leaders. If you identify and develop competent managers and supervisors, you'll ensure that you have people trained and ready to fill new leadership positions, rather than being forced to recruit unproven people externally.

Right people for the right job is the responsibility of a good manager. *Let us draw an example from the Epic Ramayana wherein Lord Rama selected the vaanar team for setting up a search operation of Sita. Ram knew very well that Hanuman, the leader of the team was capable of doing the job perfectly. As expected, Hanuman showed his power to Ravan, saw Sita, burnt parts of Lanka to show his strength and finally earned the appreciation of his boss, Ram thus motivating the whole Vaanara sena.*

Peter Senge, one of the most prominent management thinkers of our time, has quoted the Gita is his "Fifth Discipline" and "Presence." This ancient text has never been studied in the leadership context. If studied carefully, *the wisdom of the Bhagavad Gita* contains many leadership lessons that are similar to contemporary leadership theories and practices. Consider some of these lessons embedded within the Gita that are applicable to Executive level leaders of organizations.

Leaders should embrace rather than avoid formidable challenges because they bring out the leaders' greatest strengths

Leaders should be resilient in their actions and should not be weakened by pain and pleasure.

Selfish desires and animosity obscure the purpose of leadership.

Leaders achieve lasting power and glory by exercising compassion and selfless service.

Effective leaders do not lead by fear or anger.

Character is core to effective leadership.

Leaders need to be aware of the self and the surroundings.

Many contemporary leadership topics such as emotional intelligence, situational leadership, character and integrity were already discussed in the Bhagavad Gita thousands of years ago. These topics were discussed in a philosophical context, as management science as we know today did not exist then. It is also intriguing to find other management concepts embedded in the Gita. Thousands of years before Frederick W. Taylor defined work and worker, and Peter F. Drucker defined knowledge and knowledge worker, the topics of work and knowledge were already in the Bhagavad Gita.

Look for ways to train staff on a daily basis through prompt and effective feedback and offer training courses and programs to help people develop the specific skills they need. Whether you do this in-house or send people to outside training, your team should know that learning is directly connected to successful performance.

Communicate this attitude from the start. During people's induction to your team, emphasize your commitment to ongoing professional development. Encourage people to come to you with training ideas and career development plans. Make career development a strategic objective. When people can map out career paths within your team and organization, this improves staff retention and increases the likelihood that they'll develop the skills they need for the future as well as for today.

[Surah Imran: 159] "(Telling the Prophet) if you had been stern and fierce of heart they (the companions and followers) would have disappeared from around you. So pardon them and ask forgiveness for them and consult them in the conduct of affairs". This is the tip we get from Holy Quran too regarding how you should be treating people to extract maximum output from them.

In helping others, we can often get a better view of what we're capable of. It will also help you feel better about yourself. Helping others brings a wonderful sense of fulfillment and you will find yourself more confident than ever.

The owner of a company tells his employees:

"You worked very hard this year, therefore the company's profits increased dramatically. As a reward, I'm giving everyone a check for $5,000."

Thrilled, the employees gather round and hi- fi one another.

"And if you work with the same zeal next year, I'll sign those checks!"

That is something to motivate! Isn't it?

4.2 Individualization:

> *"Every word has a meaning or more than one. So talk precisely as less as possible to express your views clearly and unambiguously. Only empty vessels make noise."*
>
> *-LSS*

A project manager, hardware engineer and software engineer were in a car heading down a hill when the brakes failed. The driver managed to get it stopped by using the gears and a convenient dirt track.

All three jumped out and after peering under the car the hardware engineer said, "I see what the problem is and with this handy roll of duct tape I think I can fix it good enough to get us to the next town". The project manager quickly interrupted, "No, no, no. Before we do anything we need to decide on a vision for our future, figure out a plan and assign individual deliverables". At which point the software engineer said, "You know what, I think we should push the car back up to the top of the hill and see if it happens again."

Though this expresses the individuality of the three persons, it is not the real situation as the potentiality of different personnel is engaged in the right proportion by the project head who knows who are the members required for the project, when and how long.

There are many brilliant minds out there that have never had a chance to demonstrate what they are capable of. Perhaps a system or process could be created to discover these hidden gems and intellectual capabilities. This could lead to further development of their ideas and theories, to be used for advancement of the world. We might be in a much more advanced world than we are now.

Process individualization is usually unpredictable in nature and combines adaptive changes with the necessity of taking into account the requirements (as well as the limitations) of particular clients. Such adaptive changes usually require making immediate, comprehensive changes to existing business processes in order to align them with the clients' needs.

The war on talent has reached a turning point, if employers don't pay attention to key individuals they face the risk of losing them to other organizations

Allow leaders and high-potentials the opportunity to engage in wellness and mindfulness activities. Organizations have begun focusing on improving the work life balance of their employees. They are developing an individualized talent development plan for individuals within the talent cohorts.

People who are especially talented in the Individualization theme are intrigued with the unique qualities of each person. They have a gift for figuring out how people who are different can work together productively. Talent consists of those individuals who can make a difference to organizational performance either through their immediate contribution or, in the longer-term, by demonstrating the highest levels of potential.

Talent management is the systematic attraction, identification, development, engagement, retention and deployment of those individuals who are of particular value to an organization, either in view of their 'high potential' for the future or because they are fulfilling business/operation-critical roles.

A shrewd manager is capable of identifying the individual talents among his team members and he would use those individual talents at the behest of such requirement with no time to waste. This talent makes him a superman as he performs when others think.

Any new situation is fraught with hazards, but taking over a top job exposes a new leader to pitfalls ranging from the personal to the organizational. Accept that you can't know everything in your first six months, and even an extensive professional background can't insulate you from making mistakes in an unfamiliar company and culture. The key is to assess yourself and your progress as rigorously as you do your new colleagues and workplace, and to be prepared to make your own course corrections as you go along.

Engineers and scientists are generally seen as never making as much money as business executives will. Now, for the first time we have a mathematical proof that explains why this is in fact true.

Postulate 1: Knowledge is power.

Postulate 2: Time is money.

As every Engineer knows, Work / Time = Power

Since Knowledge = Power, and Time = Money, we have Work / Money = Knowledge

Solving for Money, we get: Work / Knowledge = Money. Thus, as Knowledge approaches zero, Money approaches infinity, regardless of the amount of work done.

Hence on a lighter note: The less you know, the more you make.

This is an outsider's view of the professionals which is not true. When work is worship, there is no dearth of satisfaction in whatever you do.

4.3 Focus and Organize:

"Your position doesn't give you the right to look down upon anyone. There is dignity in everyone and respect it and recognize it. This will help you one day in a miraculous way."

You will need to have organization skills in your life to even have a hope of paying attention to the smaller details in life. This means being organized in your work or school life, keeping track of appointments and tasks that need to be accomplished so that you aren't surprised when it's time to turn them in.

Lists are one really useful way to be organized and to make sure that you know when and how everything needs to come together. You'll be less likely to lose track of the details when you have them written out and have kept somewhere that you see every day. Have a long-term list and a short-term (weekly or daily list) so that you are able to plan for things in advance. When the items on the long-term list come up, put them on your short-term list, but this way you won't be surprised by anything in your schedule.

Once you've completed an item on your list, check it off. That way you'll know that you actually did do it and you won't be in a dither trying to remember whether you've completed each step of a certain list item.

Making a list of all pending activities is the most basic way of tracking the To-Dos and monitoring progress. There are plenty of online tools and apps these days to make this process more effective and efficient. A quick search on Google for these 'Productivity Tools' will result in hundreds of options.

Along with being organized, it is also important to create a conducive environment to focus. Many people have trouble operating effectively in dimly lit work stations with distractions. In school, a quiet corner in the library is usually a good choice. At work, do your best to personalize your desk or cubicle. I have seen people use small table fans to ventilate their space, have pictures of their family members, greeting cards made by their kids to remind them constantly of what motivates them. Changing your phone's mode to 'Silent' or even better 'Do not disturb' has worked for a few others. You need to define your own zone of focus to deliver better.

If you're working at home, avoid working in your bed and try to have a designated, organized work space. For your annoying coworker, if you can keep your door closed, do that, otherwise simply say to them "I'd love to catch up, but I really have to get this project finished. I'll come by and chat with you in a bit." Or you can tell them to go away, depending on what your relationship with them and your coworkers is like.

Multitasking spreads your attention out over a variety of items instead of focusing it onto one specific item, which means that you end up unable to give full attention to each item and you won't be able to get all the details in order.

Using the list that you've drawn up, you can go from project to project, giving each your full attention without checking your phone or social networking pages or planning what you're going to eat for dinner.

If you find yourself doing things like planning for dinner or wondering whether you've paid your bills, write down your idea or concern (you can add it to your list) and return to the project you're supposed to be working on. This way you know that you will remember to take care of the concern and you don't have to obsess over it.

Sometimes you have to multitask, or you have to conserve your energy by prioritizing the details of the project, because you have so much to do. Focus all your attention on the most important aspects so that they more of your energies.

A good way to keep your brain sharp and willing to focus on the smaller details is to allow it to take breaks. Make sure that you schedule them around the same time each day and try for 10 to 15 minutes each. This will give your brain a chance to relax in time for the next project. A break can be something as simple as stretching and taking a short walk around your office, or going for coffee down the street. When you find that you are getting really distracted or sleepy, this is a good time to find a place to do a little exercise, like jumping jacks, and to get the blood flowing. I have known people that use this time to walk around the office and catch up with their network of professional friends.

Including meditation in your routine is something that has been proven to be incredibly useful for a wide variety of things. It not only helps you with your physical and mental health but also with your memory and your attention to detail, by calming your mind and reducing stress (it helps put your brain on more positive neuro-pathways).

Find somewhere quiet to sit for about 15 minutes each day (when you're more advanced at meditation you can do it anywhere: at your desk at work, on the bus, etc. but it's good to start somewhere without many distractions). Close your eyes and take deep breaths all the way down into your belly. Focus on your breathing. When you find distracting thoughts coming into your head, acknowledge them but do not focus on them. Return to your breathing by saying to yourself "Breathing in, breathing out."

If you're struggling to stay focused on the big project that's in front of you—because there are a million other "little things" that need to get done - do a 'brain-dump' before you sit down at your desk. Make a list of everything that's distracting you - things to remember, things to research later, ideas you don't want to forget - and save your list for later.

Then, sit down and focus on the big task at hand - knowing that all of the "other stuff" is safe and secure, preserved on your list for later. Making a brain-dump list is also a smart thing to do at night, just before you fall asleep - especially if you're plagued with a "chattering mind" that won't quiet down at bedtime.

Among all the substances we misuse and abuse, the greatest is time. Time is life; we squander it at our peril. Killing time deadens us. Actually, it's not time we lack; it is focus, awareness and a sense of priorities. We must change the space of the pace – wake ourselves up by shifting to another way of being. We have all the time in the world. It's up to us to choose how to use it.

Lama Surya Das in his book, "Buddha Standard Time: Awakening to the Infinite Possibilities of now," talks about Time Management which is very essential for any management as "Time-sickness" is rampant today. People say they want to slow down and live more naturally and in a healthy and sane manner, but who knows how to actually do so, has time-medicine available, and is also ready, willing, and able? Awareness is the essential ingredient in this great journey, delivering to us the bigger picture as well as minute details along the way.

4.4 Deliberative:

> *"Good management consists in showing average people how to do the work of superior people."*
>
> *– John Rockefeller*

Dialogue and deliberation are creative processes that help people come together across differences to tackle our most challenging problems. In a time of extreme political partisanship and increased conflict between religious and ethnic groups, teaching, spreading, and supporting the skills of dialogue and deliberation are vital.

Deliberative dialogue is a form of discussion aimed at finding the best course of action. Deliberative questions take the form "What should we do?" The purpose is not so much to solve a problem or resolve an issue as to explore the most promising avenues for action. Following a usage that traces back to the ancient Greeks, deliberation can be defined as the process of establishing intent and resolve, where a person or group explores different solutions before settling on a specific course of action. "We deliberate not about ends," said Aristotle, "but about the means to attain ends." Deliberation is necessary for what is uncertain, he noted, when there may be reasons for deciding on one course of action but equally compelling reasons for deciding on another.

Deliberation is an approach to decision-making in which citizens consider relevant facts from multiple points of view, converse with one another to think critically about options before them and enlarge their perspectives, opinions, and understandings.

At the most basic level, deliberation is what should occur before a decision has been made and change management is what needs to occur after. This formula, however, only scratches the surface of how the two fields can enrich each other, and as yet there is a dearth of shared knowledge between them.

"Deliberation" is defined as the critical examination of an issue involving the weighing of reasons for and against a course of action. Deliberation can involve a single individual, but the deliberative processes under discussion involve group deliberation. Thus, we define a "deliberative process" as a process allowing a group of actors to receive and exchange information, to critically examine an issue, and to come to an agreement which will inform decision making.

The concept of deliberation in the present use depicts the exchange of reasons. It portrays the ordinary process of reaching collective decisions by giving and taking reasons. Arguing or argumentation depicts a principled form of deliberation; a procedure for redeeming validity claims – for testing justice claims as well as of truth claims in a rational discourse.

In policy-making deliberation is generally needed

- To coordinate actions, as without deliberation interdependent actors will not be
cognizant of problematic situations;
- To solve problems rationally because without arguing we do not know the Counter arguments;
- To resolve conflicts over distributive shares legitimately, because without deliberation
we do not know whether all viewpoints are heard and given due consideration;
- Deliberative inquiries held to enlighten the actors, shed new light on the issues at stake, and even change actors' attitudes or beliefs if they can be.

The modern world realized the importance of joint consultation (Shura) and team-work when the Japanese based its management style on it and proved to the world of its effectiveness.

The Holy Quran advocates this concept:

[42:83]... and those who do their work through mutual consultation.'

[3:159] Pass over (their faults), and ask for (God's) forgiveness for them; and consult them in affairs (of moment). Then, when you have taken a decision put your trust in God, for God loves those who put their trust (in Him).'

There are enough examples of this theory in Ramayana too regarding consulting subordinates on important matters and allow them to give their opinions freely, which we call in management terms as "Brain Storming".

When Vibhishan defected from Ravana, Rama took him under his protection. He then had a talk with the various army chiefs some of whom disagreed with Rama. Instead of punishing them, Rama assuaged their suspicions and got them to accept his decision. Everybody felt that their opinions had been heard and that their objections had been clarified. Empowerment of subordinates to question his decisions was a key and unique quality of Rama which one cannot but help comparing with Ravana who never allowed anybody to contradict him.

Deliberating is the most powerful and successful tool to bring people together and to address various tickling issues and to get over on top of them.

The possible solutions to a given problem emerge as the leaves of a tree, each node representing a point of deliberation and decision. -Niklaus Wirth

What do we tell our children? Haste makes waste. Look before you leap. Stop and think. Don't judge a book by its cover. We believe that we are always better off gathering as much information as possible and spending as much time as possible in deliberation. –Malcolm Gladwell

4.5 Command:

"A leader is the one who can outline the broad vision and the direction, and say here's where we are going to go, here's why we need to go there, and here's how we are going to get there. A manager is the one who actually gets up under the hood and tunes the carburetor."

- Mike Huckabee

To be the leader of an organization and critical projects, the top level executives need to have a strong hold on the following:

-Appoint the right project manager/s for the job

-Support the project manager with the right team.

-Understand the strengths and weaknesses of the project team

-Prioritize tasks and come up with guidelines for when priorities conflict.

-Actively monitor projects, as well as your team progress (learning, team building etc.)

-Hold periodic meetings and review progress effectively

-Manage change happening around due to environment

-Take a hard line against scope creep

-Create milestones for every member of the team—and celebrate them when met

-Keep track of the time spent by the key personnel responsible for the completion of major parts of the project

You can be a good commander only if you know how to obey. One can become a good commander by going through the ladder step by step and understanding each step clearly and remembering them.

> *"All that is necessary to break the spell of inertia and frustration is this: Act as if it were impossible to fail. That is the talisman, the formula, the **command** of right about face which turns us from failure to success." -Dorothea Brande*

"To **command** is to serve, nothing more and nothing less."--- Andre Malraux

Your leadership and command will have value only if you have proven yourself time and again that you meant success every time and achieved your goal.

Create an unshakeable confidence on you from your people which you can achieve only through your performance.

The command may be in the form of your expression, your communication skills, your convincing skills, the way you hold others attention with your genuine attitude and sincerity and perseverance. The more you rise in the ladder, the more polite and humble you must be which in turn adds more sincerity to you as commander.

A true commander is the one who installs a great amount of confidence in his team, whatever the combination of the team. The best example we draw from Ramayana, the great epic of India describes about this.

Rama led what was essentially a rag-tag army against the sophisticated army of Ravana. The Rakshasa army was a powerful one, which had defeated the formidable, devas and vanquished powerful kings. In contrast, the army of Rama comprised of soldiers who were perhaps aboriginal tribes who had never encountered a sophisticated army before. Not surprisingly, Ravana and his courtiers jeered at the army and laughed scornfully at Angad, Ram's messenger who had come with an offer of peace. Yet Rama maintained confidence in the ability of his army to surmount this seemingly impossible odd and enthused by his confidence his army fought to achieve victory.

Set ambitious goals and motivate troops to meet them.

Commanding is not everybody's cup of tea. It requires understanding, realization and purpose, sensible, not coupled with egoism and proudness.

There was a King ruling the territory of Amarapoora. He was very much loved and liked by his people and they were proud of their king for his valor, intelligence, shrewdness, his gigantic figure etc.

Whatever he wished were readily available for him. Whatever he prayed to God was sanctioned to him.

He had supreme power vested in his hands, but he was indifferent to it. He owned everything that his heart could desire, and those very possessions were killing him. He was getting restless and fatigueness started setting on to him.

There were those who travelled especially to other countries, but to return and tell him of all that they had seen, and of how inferior all lands and rulers were when compared with their own.

And, like a child, he was craving for something new desperately.

One fine morning, one person from India reached the kingdom and expressed his interest to meet the king and also passed on the message that he had some new possession to be shared with the king which would make the boredom of the king be gone. He was asked about the details of his proposal. He explained that he had brought in a new game called "Chess" from India which would make the king ever engaged with fullest interest. Finally, the king's permission was granted to him and he got an appointment with the king to demonstrate the new game. As he proceeded with his demonstration, the king got more and more enthused and involved and finally he won the heart of the king. Day in and day out, the game was spreading all around and the Indian became the chief guest of the kingdom. The growing popularity of the Indian was not liked by some in the court and so to demean him, they were on the lookout for an alternative.

Finally, they could locate one by name Nicomar from a distant land, who was great in painting cards and different dices to play which were unique and never seen before for them. He was taken to the king who was always in the lookout for a change. Slowly, his painting and dice play started attracting the king and simultaneously the importance of the Indian started fading.

Soon, the king got fed up with Nicomar too and added penalties to every small mistakes he did.

One fine day, the king called him and said, "Nicomar, I am feeling bored to see the sea around my kingdom filled with salt water. I want you to fill it up with milk which will give a new look and respect to my kingdom."

Nicomar was thunder struck and said," Your Highness! How is it possible?"

"Nothing is impossible in my territory. I hate to hear anyone telling that something is impossible. This is my order. You shall fill up the sea here with milk. I give you fifteen days time for the task. If you fail, you shall be beheaded."

Nicomar was crest fallen and started repenting every moment for his entry into this kingdom. He could never eat or sleep properly. He was cursing at his fate every day. By the end of fourteenth day, he happened to meet the Indian who was also living around the vicinity by chance. He told the Indian, "My dear friend, my arrival here had made you insignificant, but today I stand most insignificant and tomorrow I am be gone."

The Indian called him close by and whispered something in his ears, Nicomar's face got brightened suddenly. He embraced the Indian and left to his home. Next day, he was taken to the court hall of the king and everyone was anxiously waiting to see what was going to come.

The king asked Nicomar, "So, are you ready Nicomar?"

Nicomar, "Your Highness! May I confirm what your order to me was first?"

"Yes, go ahead", said the king.

"I am to fill the sea around your territory with milk. Is that right your Highness?"

"Yes, absolutely. Are you ready?" asked the king.

"I am ready Your Highness!" said Nicomar.

"Then go ahead." Said the king.

"Your Highness! My task is to fill the sea with milk and I am ready to execute it. But, the sea is now full of salt water. Kindly arrange to empty the water from the sea so that I can fill milk into it."

The whole court room was dumb struck hearing this and they were expecting a major quake from the king.

The king after a while of thoughtful mood, turned towards Nicomar and smiled at him with a sign of acknowledgement of his intelligence and shrewdness.

"Nicomar, You have proved your greatness. I am very much pleased on your performance. Tell me what gift you want?" said the king.

"Your Highness! I am happy that I am with my head in its place today and I am not losing it. I fully dedicate this to my dear Indian friend who came to my rescue by giving me the solution to my problem", said Nicomar.

The king in appreciation of their crisis management ability rewarded both of them and treated them as special guests for some more days till they wanted to take leave of his kingdom.

One who is able to command the situation rightly even at the time of crisis will definitely emerge as a very successful manager.

4.6 Responsible:

> *"Do not think that you are always correct. Your decisions might be based on assumptions which might prove wrong. Be diligent."*

Leaders are responsible for "getting stuff done," but let's break it down. Leaders are responsible for ensuring the following are true:

* All employees are doing their jobs correctly, thoroughly, and on time
* Expectations and goals are clear
* Conflicting priorities are addressed and readjusted as needed
* Objectives and goals are being met or exceeded
* Key information is conveyed up the ladder, to the leader's manager or others who might need to know
* Employees are given a level of oversight appropriate to their position and abilities
* Good employees feel appreciated, heard, and as if someone is "looking out" for them
* Employees are given regular feedback about their performance, including what they do well and where they need to improve, with special attention toward low performers to ensure they improve or are transitioned out
* Staffers are representing the company and department appropriately to the public and various stakeholders
* Employees are following company policies
* There is a plan in place to ensure continuity if disaster were to strike (for instance, if a key staffer were to disappear tomorrow, is there a way for you to access important documents, and other information someone would need to step in?)

And, finally, and hugely important, managers are responsible for ensuring results in their realms-concrete, measurable results.

The leader should take complete responsibility in making sure his managers are able to oversee, or administer the operation of the area that they are responsible for. The day-to-day requirements vary, and the leader must be able to size and alter the task at hand, sometimes on a moment's notice. Most operations also include reporting and financial requirements, as well as personnel and legal obligations that must be met on a timely basis. The manager must be able to ensure that all requirements of his department are met on time.

The operation of a department or organization usually requires more work than any one person could get done on his own. A great leader is one that is able to delegate workload effectively and monitor its progress towards completion while not delegating the ultimate accountability. The leader must also know exactly what each of his staff is capable of and give them work that they can complete effectively while also challenging them to achieve more.

A manager must be proficient in a number of areas to be an effective leader, one who can motivate employees to perform at their highest capabilities. A leader differs from his subordinates because he must measure his success by what he can get others to accomplish, and not solely by what he can do on his own. While opinions vary about a leader's specific top responsibilities, performing certain key functions effectively will ensure long-term success in management.

Formulating policy is one of the core duties of an operations leader. Companies must operate and function on a daily basis within a prescribed set of guidelines. These guidelines are generally established by operations leaders. These can include how different departments within the company or organization communicate and cooperate with one another. Policies can also include disciplinary actions taken when employees break company rules.

An important and core responsibility of an operations leader is communicating with other management professionals within the organization to keep the company running smoothly, and communicating with other companies and organizations with which the company does business.

The leader's task will always remain getting the work done, but this doesn't mean getting it done by any means possible or at any cost. The leader who finds ways to improve the operation with more effective methods is making as important a contribution to success as he or she can make by achieving quotas and meeting schedules.

There are a number of steps a leader can take to eliminate waste and inefficiency. These include identifying lost time, coordinating tasks and resources, improving workflow, improving scheduling and eliminating inefficiencies in the use of employees.

There should be goal setting without which there is no direction to progress. In Bible, this sort of goal setting in management is referred to in Proverbs 21:5, as below:

"The plans of the diligent lead surely to plenty,
But those of everyone who is hasty, surely to poverty."

However small or big be the task entrusted, one should take it serious and feel responsible for accomplishing the task successfully. Following is a short story by Pedro Pablo Sacristan illustrating this aspect nicely.

*The day when the jobs were handed out was one of the most exciting for all the children in the class. It took place during the first week of the term. On that day, **every boy and girl was given a job for which they would be responsible for the rest of that school year**.*

*As with everything, some jobs were more interesting than others, and the children were eager to be given one of the best ones. When giving them out, **the teacher took into account which pupils had been most responsible during the previous year**, and those children were the ones who most looked forward to this day. Among them Rita stood out. She was a kind and quiet girl; and during the previous year she had carried out the teacher's instructions perfectly. All the children knew Rita was the favorite to be given the best job of all: to look after the class dog.*

But that year there was a big surprise**. Each child received one of the normal jobs, like preparing the books or the radio for the lessons, telling the time, cleaning the blackboard, or looking after one of the pets. **But Rita's job was very different**. She was given a little box containing some sand and one ant. And even though the teacher insisted that this ant was a very special ant, Rita could not help feeling disappointed. **Most of her classmates felt sorry for her**. They sympathized with her, and remarked at how unfair it was that she had been given that job. Even her father became very angry with the teacher, and, as an act of protest, **he encouraged Rita to pay no attention to this

insignificant pet. However, Rita, who liked her teacher very much, preferred to show the teacher her error by doing something special with that job of such little interest.

*"**I will turn this little task into something great**," Rita said to herself.*

*So it was that Rita started investigating all about her little ant. She learned about the different species, **and studied everything about their habitats and behavior**. She modified the little box to make it perfect for the ant. Rita gave the ant the very best food, and it ended up growing quite a bit bigger than anyone had expected...*

*One day in spring, **when they were in the classroom**, the door opened, revealing a man who looked rather important. The teacher interrupted the class with great joy, and said,*

*"**This is Doctor Martinez**. He has come to tell us a wonderful piece of news, isn't that right?"*

*"Exactly". Said the Doctor. "**Today they have published the results of the competition**, and this class has been chosen to accompany me, this summer, on a journey to the tropical rainforest, where we will be investigating all kinds of insects. **Among all the schools of this region**, without doubt it is this one which has best cared for the delicate little ant given to you. Congratulations! You will be wonderful assistants!"*

*That day the school was filled with joy and celebration. **Everyone congratulated the teacher for thinking of entering them in the competition** and they thanked Rita for having been so patient and responsible. And so it was that many children learnt that to be given the most important tasks you have to know how to be responsible even in what are apparently the smallest tasks. And without doubt, it was Rita who was most pleased at this, having said to herself so many times "I will turn this little job into something really great".*

Another interesting story on assuming responsibility goes thus:

Two sons work for their father on the family's farm. The younger brother had for some years been given more responsibility and reward, and one day the older brother asks his father to explain why?

The father says, "First, go to the Kelly's Farm and see if they have any Geese for sale – we need to add to our stock." The older son soon returns with the answer, "Yes they have five Geese, which they are ready to sell to us."

The father then says, "Good, please ask them the price."

The older son returns with the answer, "Each Geese cost us $10."

The father says, "Good, now ask if they can deliver the Geese tomorrow."

And duly the older son returns with the answer, "Yes, they can deliver the Geese by tomorrow."

The father asks the older son to wait and listen, and then calls his younger son from a nearby field, "Go to the Davidson's Farm and see if they have any Geese for sale – we need to add to our stock."

The younger son soon returns with the answer, "Yes, they have five Geese for $10 each, or ten Geese for $8 each; and they can deliver them tomorrow – I asked them to deliver the five unless they heard otherwise from us in the next hour. And I agreed that if we want the extra five geese we could buy them at $6 each."

The father turned to the older son, who nodded his head in appreciation. He now realized why his younger brother was given more responsibility and reward.

A lot of times we face the question why someone is so successful and someone is not. Some people call it luck, some people call it being in the right place at the right time and some people call it influence, however the truth is performance with full responsibility and full potential. These two aspects are constantly observed by all and this is what ultimately leads to success of any individual.

Responsibility is taking care of your duties. Responsibility is answering for your actions. Responsibility is accountability. Responsibility is trustworthiness.

4.7 Strategic:

> **"Expression is to be genuine and need not be dramatic unless it is called for."**

"The essence of strategy is choosing what not to do." Says Michael E. Porter

Strategic leaders focus on achieving business results. Operationally oriented projects tend focus more on just getting the work completed. By focusing on improving customer satisfaction, beating the competition and analyzing market data, strategic project managers ensure long-term success and profitability. Instead of focusing on short-term results, such as meeting deadlines and operating within the budget, strategic project managers have a long-term perspective. They ensure that their project's goals align with the company's strategic mission and objectives.

Experienced strategic project managers typically have responsibility for mentoring less experienced project managers. By sharing tips, techniques and templates, they help others learn how to perform this role. You work closely with team members to provide training on strategic initiatives, such as improving quality. You also typically facilitate meetings with sponsors and stakeholders to identify project requirements and ensure they align with your company's strategic goals.

Strategic thinking is a process that defines the manner in which people think about, assess, view, and create the future for themselves and others. Strategic thinking is an extremely effective and valuable tool. One can apply strategic thinking to arrive at decisions that can be related to your work or personal life. Strategic thinking involves developing an entire set of critical skills. What are those critical skills?

Critical Skill 1: Strategic thinkers have the ability to use the left (logical) and right (creative) sides of their brain. This skill takes practice as well as confidence and can be tremendously valuable.

Critical Skill 2: They have the ability to develop a clearly defined and focused business vision and personal vision. They are skilled at both thinking with a strategic purpose as well as creating a visioning process. They have both skills and they use them to complement each other.

Critical Skill 3: They have the ability to clearly define their objectives and develop a strategic action plan with each objective broken down into tasks and each task having a list of needed resources and a specific timeline.

Critical Skill 4: They have the ability to design flexibility into their plans by creating some benchmarks in their thinking to review progress. Then they use those benchmarks to as a guide and to recognize the opportunity to revise their plans as needed. They have an innate ability to be proactive and anticipate change, rather than being reactive to changes after they occur.

Critical Skill 5: They are amazingly aware and perceptive. They will recognize internal and external clues, often subtle, to help guide future direction and realize opportunities for them and their companies or organizations. Great strategic thinkers will listen, hear and understand what is said and will read and observe whatever they can so that they will have very helpful and strategic information to guide them. Strategic thinkers often have those "Ah Ha" experiences while on vacation, walking, sitting and relaxing or during many other activities because they see or hear something that resonates and because they are so aware and perceptive.

Critical Skill 6: They are committed lifelong learners and learn from each of their experiences. They use their experiences to enable them to think better on strategic issues.

Critical Skill 7: The best and greatest strategic thinkers take time out for themselves. Their time out may be in the form of a retreat (some prefer to call it an "advance" since it "advances" their thinking"); a walk in a special environment; relaxing in a comfortable chair in the lobby of an historic hotel; or an afternoon in a quiet place with a blank sheet of paper or their lap top computer with "their thinking caps" on.

Critical Skill 8: They are committed to and seek advice from others. They may use a coach, a mentor, a peer advisory group or some other group that they can confide in and offer up ideas for feedback.

Critical Skill 9: They have the ability to balance their tremendous amount of creativity with a sense of realism and honesty about what is achievable in the longer term. This ability to balance does not deter them in their thinking. Sometimes they refer to themselves as realistic optimists.

Critical Skill 10: They have the ability to be non-judgmental and they do not allow themselves to be held back or restricted by judging their own thinking or the thinking of others when ideas are initially being developed and shared. This is especially true during any brainstorming exercises to ensure a flow of great ideas. There will be time to test the ideas AFTER the brainstorming is concluded.

Critical Skill 11: They have the ability to be patient and to not rush to conclusions and judgments. Great ideas and thoughts require time to develop into great successes in the future to reach your defined vision.

To Think Strategically, First Think Critically. Before there is strategic thinking, there must first be a strong foundation of critical thinking. Strategic thinking is the ability to see the total enterprise, to spot the trends and understand the competitive landscape, to see where the business needs to go and to lead it into the future. That's the top level of the pyramid. But like all pyramids, it needs a solid base.

"The real challenge in crafting strategy lies in detecting subtle discontinuities that may undermine a business in the future. And for that there is no technique, no program, just a sharp mind in touch with the situation" says Henry Mintzberg

At the foundation it is self-awareness, which arises from the ability to think critically, along with an intellectual openness. Those, in turn, provide the basic business skills – including decision-making, problem-solving, and a solid grasp of both the business and the customers it serves.

The higher one rises in an organization, the more these skills are needed. And as they are developed, they build the next layer up – the ability to embrace change and ambiguity, and in so doing create something new and different. That, in essence, is strategic thinking.

But it all starts with critical thinking. And those who can't master the basics don't go far in an organization.

A new manager spends a week at his new office with the manager he is replacing. On the last day, the departing manager tells him, "I have left three numbered envelopes in the desk drawer. Open an envelope if you encounter a crisis you can't solve."

Three months down the road there is major drama in the office and the manager feels very threatened by it all. He remembers the parting words of his predecessor and opens the first envelope. The message inside says "Blame your predecessor!" He does this and gets off the hook.

About half a year later, the company is experiencing a dip in sales, combined with serious product problems. The manager quickly opens the second envelope. The message read, "Reorganize!" He starts to reorganize and the company quickly rebounds.

Three months later, at his next crisis, he opens the third envelope. The message inside says "Prepare three envelopes."

The important of strategic planning precisely with time management can be seen from Mahabharat.

It is well known that the great battle of Mahabharat was the consequence of a long series of intrigues and conspiracies planned and abetted by the wicked prince Duryodhana, the eldest son of Dhritarashtra. Right from the very beginning, it had been a major (and the only) preoccupation of Duryodhana and his maternal uncle, Shakuni, to deprive the Pandav brothers of their legitimate rights. In order to achieve their goals, they were bent upon using any methods foul or fair. The duo did not let any opportunity slip by to eliminate Pandavas or at least to keep them away from the throne of Indraprastha.

One such occasion for hatching a conspiracy presented itself when the Pandavas were sent to Varnavrat festival. Unknown to them, Duryodhana and Shakuni had an apparently magnificent palace erected for the Pandavas. The grandeur of the palace impressed the Pandavas so much that they felt a very deep sense of gratitude towards their cousin Duryodhana who, in their view, took so much pain to make their stay at Varnavrat so comfortable.

However, Vidur, the uncle of the Pandavas and Kauravas, was not convinced that Duryodahana and Shakuni could be doing all this as a matter of affectionate hospitality. Vidur has always been admired – and rightly so – for his wisdom. He viewed and examined everything critically;

135

more so where there was involvement of Duryodhana or Shakuni. He found out that the apparently grand palace was built of wax.

And why wax? Shakuni had worked out his scheme with meticulous planning. He had planned that the Pandavas would be invited to Varnavrat festival and lodged in this palace of wax, where they would be provided with generous hospitality. Their needs would be attended to with utmost care and politeness. In the process, they would develop an uncritical and affectionate attitude towards Duryodhana. Nothing would be done which might incite their suspicion. They would thus stay till the last day of the Varnavrat festival. Doing anything untoward during the days of the festival would create confusion and invite the ire of the general population who are joyfully participating in the festival.

Once the festival was over and the Pandavas were left alone in the palace, they would be requested as a matter of courtesy to stay on for a couple of days more. On the last night of their stay, the palace of wax would be put on fire. The unsuspecting Pandavas would be consumed by fire in their sleep. The world would know of the accident only to sympathise with Duryodhana rather despise him.

As indicated above, Vidur had anticipated some trouble, and could read through the wicked designs of Shakuni and Duryodhana. By the time he got to know the totality of the conspiracy, only one week was left. But he lost no time in arranging to do what the situation demanded. He sent for a very skilled Khanik (tunnel maker) and asked him to prepare a tunnel well before the night when the wax palace was to be put on fire. In the meantime, Vidur was able to send across his message to the Pandavas as well, indicating to them as to how they would escape from the palace without the knowledge of the men of Duryodhana, who were all around the palace on a constant vigil.

As pre-planned, the wax palace was put on fire and with the help of foresight of Vidur the tunnel was completed in time and all the Pandavas walked out of the palace, untouched by the evil designs of Duryodhana and Shakuni.

This explains how contingency plans are essential in strategy planning.

4.8 My Experiences as Executive:

> *"If you have a dream, don't just sit there. Gather courage to believe that you can succeed and leave no stone unturned to make it a reality."*
>
> *—Roopleen.*

Having had a view on the essential characteristics of an Executive in Management let me share my personal experiences which demanded these characteristics. This is a difficult position which requires knowledge, quick thinking, team spirit, cajoling, consoling, guiding and all sorts of emotional and technical and administerial tensions one has to handle in a cool and composed manner. I have learnt them through my experience and interactions with my team as well as my bosses. This is the position where you have to handle effectively your subordinates as well as your bosses without antagonizing anybody.

As far as I am concerned, this was the difficult phase of my career. So long, as a Novice and Middle Level Manager my contribution was as a sincere, obedient subordinate to my boss. Of course, participated in the brain storming sessions with higher officials where the decision making ultimately was with the bosses only. Yes, many a time, my suggestions have been taken while sometimes not. But the phase of my service as Executive Level Manager only taught me several intricacies of project and personnel management. It transformed me from innocence (not ignorant) to intelligence, from obedience to obedience as well as commanding, from waiting for decisions to decide and communicate, more conscious of time, quality and cost.

As this level is in between the top and bottom, managing both ends require special skills as the expectation of both sides are to be met tactically by the executive level managers. This is the position where one has to take the blame for the mistakes if any committed by both sides while the achievements will be the credit to top level management. In short, this is the position one can be easily made a scape goat. My experiences at this phase, both good and bad had made me to think differently to succeed every time, sometimes easily and sometimes with hard and intelligent maneuvering. This level of management experience makes everyone a well seasoned executive and make them fit to become a top level manager.

4.8.1 An Introduction to the Tsunami:

While I served as Deputy Chief Engineer (DCE) at Port Blair, I was second in command in the department to the Chief Engineer and Administrator (CEA).

The 26th December, 2004! Yes! It was an ominous day to the Eastern part of the world, when a monster earthquake with an intensity of 9.2 on the Richter scale with its epicenter at Banda Ache of Indonesia, rocked this region during the wee hours of the morning resulting in giant tsunami waves lashing not only the coast of Indonesia, but also the Andaman and Nicobar Islands due to their geographic proximity.

The intensity of tremors felt in Andaman and Nicobar Islands were of the order of 8.6 on the seismic Richter scale. The havoc caused by the recurring tsunami waves in these islands was horrendous, killing thousands of people and damaging everything on its way. The waves were reported to be of the order of 6 to 8 meters at different locations of the islands. Almost all the harbor structures existing in various parts of the islands got damaged either partially of fully. As shipping was and is the lifeline of these islands, the urgency of rehabilitating the port structures was on top most priority and hence the pressure on our department started mounting sky high. It was the time for dynamic leaders across the islands to abundantly prove their capabilities by rising up to the occasion and tackling the challenge that Mother Nature had created for them. My organization's CEA was on his way to retirement on superannuation in the next few months. Keeping this back ground in mind, I would like to share with you a very important chapter of my life!

You all must be aware that the tsunami damages were first streaming in all news channels projecting Car Nicobar which is the headquarters of Nicobar District, where the office of Deputy Commissioner of A & N Administration was located. This island basically has a flat terrain. The sea water in the form of tsunami waves entered through one side of the island and left the other side in many islands, damaging acres of vegetation, innumerable coconut trees, buildings, defense establishment installations, their family accommodation buildings and not even sparing all wells which served as potable water sources. All power lines too got destroyed adding to the misery. There was nothing to eat and nothing to drink except sea water.

Such was the absolute emergency situation where rehabilitation package including food and water had to be transported there immediately!

4.8.2 Emergency Rehabilitation Management at Mus Harbor, Car Nicobar:

> *"Avoid giving advice. It is a tool to be used only and only if it is asked for."*

> *-LSS*

The main problem was that though Mus Harbor of Car Nicobar had only minor damages to the break water, the berthing structure was safe and was ready to take ships. However, the approach to the berthing wharf got subsided by 40 cm during the severe earth quake which posed a great challenge. This made berthing of sea vessels almost impossible which implied the rehabilitation package could not be made available immediately as the rehabilitation materials could not be transported on to the land area because of the situation that no vehicle could pass out from the berthing area to the shore because of this subsidence of approach.

The Lieutenant Governor (LG) who was the head of the state, commissioned the CEA of ALHW. In turn, CEA requested me to join him to meet LG. He requested CEA to look into the emergency situation of Car Nicobar and take appropriate action most urgently on war footing basis. He also suggested that CEA could visit Car Nicobar the very next morning as the situation was very demanding. My boss assured the LG that since his presence was required at Port Blair to coordinate all rehabilitation activities, his senior most DCE, i.e. me would proceed to Car Nicobar the next day, if the LG wishes so.

LG was visibly happy at his proposal and said to me, "Mr. Sekar, please proceed in our Air force plane tomorrow morning itself. My office will make all arrangements. From Carnicobar field you will be air winched at Mus Harbour." I agreed immediately. The gravity of the situation was palpable in our meeting room.

Promptly, the next morning, i.e., 27th, December.2004, I was carried by an air force carrier to Carnicobar air field. When I reported to the air field in charge, he promptly replied, "Yes Sir, we have received instructions from office of LG to drop you at Mus. Please wait for about half an hour by which time your helicopter will be ready."

During the wait for my ride, I thought to myself that I would take a scientific approach and inspect the damage caused by the tsunami in the operational defense area, where the damage was very heavy. As I was strolling in the area, I could see almost all the residential and office buildings torn down mercilessly, several televisions, scooters, two wheelers strewn haphazardly all over in damaged conditions, the main power house totally blown out with big holes on the walls and looked like an ancient town marred in a world war.

As I traversed inside further, I could see lots of pegs with red flags fluttering in the breeze. I couldn't understand what they were till I turned the other side! My God! A band of people were engaged in removing the dead bodies of the victims of tsunami from the debris and after removing each body, they planted a peg with a red flag on top of it, which gave the count and locations of the dead bodies recovered. That moment was the most horrific I have ever experienced in my life! I could not take it any more as my lungs were chocking.

I moved away from the location and started moving towards the sea. When I was about 50 meters away from the beach, I saw a big steel barge lying, which was to be used for transporting cargo between ship and the shore and it was to be in floating in sea. I was surprised to see it lying at that distance from the coast. When I was wondering, a few more people came near me and casually I was expressing my observation to them, "See, There are so many coconut trees standing between the beach line and this location. And there is no gap between them for such a big barge to pass through. I am wondering as to how this barge was brought here!" Immediately, one local Nicobari young man who was hearing me shot back, "Sir, It was not brought here. It was hauled from the sea to this place by the first tsunami wave!" I was stunned to hear this. With doubt I asked him again, "This is too heavy, may be of the order of about 60 Tons. How is that possible that such heavy barge could be lifted and carried all the way?" He replied again, "Sir, on that day in the morning, I was sitting on this barge and fishing with my fish line. All of a sudden, the barge went deeper into sea. When I was wondering what was happening, I was flying in the sky along with the barge in the tsunami wave over the coconut trees and landed with a thud here. I ran helter-skelter into the land and saved myself from washing away in the return of the wave. I shudder to think of it even!" I could see him trembling while he was narrating the event! That was the effect of tsunami.

I came back to air field where the helicopter was ready. I was taken in the chopper immediately and after flying around 15 minutes, I was winched down at Mus (as there was no helipad landing at Mus). Though Mus was earlier connected by road, all roads were damaged by tsunami waves and most of the uprooted coconut trees were lying all around thereby making to Mus by road surface was impossible.

When I landed in Mus, I could see a sigh of relief ion the faces of my department staff. My officers there wereS/Shri. Mr. Sankaran Assistant Engineer (Civil), Mr. Murugaboopathy, Inspector of Works (Mech.), Mr. PremjiJohn, JuniorEngineer (Civil) along with field staff. Firstly I reassured to them that we can bring the situation under control and motivated them by saying that, that was the time to prove ourselves to the local population and to the state.

We went straight to the site of damage and started observing the extent of damage and what could be done urgently to ensure movement of vehicles from the berth to land. By then so many people gathered to see what we were going to do. Several suggestions started flowing from several people like arranging to fill the sunken depth with concrete, fill the sunken portion by transporting earth from inland, spread stone boulders to raise the height etc. After seeing the site situation, availability of men and materials, I decided to collect the concrete rubble lying all around from the damaged buildings, transport them over the minimum distance from the gate to the site manually making use of all manpower available with us and also utilizing voluntary services and dump the rubble to form an easy smooth ramp to enable vehicles to negotiate the difference in depth over a mild slope. This was the easiest and fastest solution possible at the moment and we started executing it immediately. Several people around watching this offered their services voluntarily which helped us to complete our mission by 4 pm the same day. Immediately a message was relayed through police wireless to LG and Administration "Mus harbor rehabilitated temporarily. Ready to receive ships in harbor".

Next morning, when we received a ship with rehabilitation materials, the amount of satisfaction we all had was tremendous and unexplainable. We all realized that "Service to Humanity is service to God"

4.8.3 Need For In-depth Knowledge on the Project:

"Withhold not good from them to whom it is due,
when it is in the power of thy hand to do it."

There were a lot of spots in Andaman and Nicobar Islands where there had been lots of sea erosion observed along the shore line and the local population had started demanding for immediately preventing this failing which a large area of usable land would be lost to sea. This was discussed at length in the chamber of the Chief Secretary (CS) of A &N islands at Port Blair. Mr. Shakti Sinha,I.A,S., a dynamic officer was the Chief Secretary then. There was one Superintending Engineer of Andaman Public Works Department (APWD), which was the main construction department of Andaman Administration present in the meeting. I was representing my department of Andaman Lakshadweep Harbour Works (ALHW). During the discussions, this gentleman started explaining the CS that he had the experience of construction of shore protection wall elsewhere and he could organize the same in A and N islands too.

As CS knew that I specialized in Ports and Harbor structures and the shore protection activities that fell under that category, C.S turned towards me with a look if I wanted to say something. I said, "Good if he knew about shore protection, Sir."

CS to SE (APWD): How do you do it?

SE (APWD): Sir, I have done a sea wall construction four years back at …location and I still have the details. I can do the same at these locations too without any problem.

CS to me: Mr. Sekar, would you like to see his details some time?

Me: No sir, I have a few questions for him.

CS: Go ahead.

Me to SE: Are you proposing to use the same design and drawings to all the other locations too?

SE: Of course, yes. Those design and drawings were proved successful.

Me: Perfectly all right. Do you have an idea of how those design and drawings were made?

SE: I don't know about it. My department supplied the drawings. I executed it.

CS to Me: Sekar, if you have something to say, come out.

Me: May I explain to him with a small example, Sir?

CS: Go ahead.

Me: There were two couples living side by side in an apartment building, one couple was old and the other, newly married. One night, the old lady developed stomach pain. The old man knocked at the door of the young couple and the young man opened the door and asked what the matter was. When the old man explained, the young man ran to a nearby doctor and brought him home. The doctor after checking her condition gave a prescription to cure her stomach pain. The young man went to the nearby pharmacy and brought the medicines and administered to the old lady. She became alright soon and thanked the boy. After a few days, the young lady, the wife of the young man developed stomach pain. The young boy ran to the medical shop with the same prescription given by the doctor to the old lady and brought the same medicines and administered to her. Within a few minutes, the young lady started crying with pain more and more. Then he ran and brought in the doctor who was extremely angry at his action.

Every situation is unique. The shore protection at any particular place is wholly dependent on site specific data collected like beach slope, direction of approaching waves, intensity of waves, grain size distribution of the local sand material at the location, maximum height and periodicity of approaching waves etc. and after collecting all such details, a model study is suggested to design the foundation and sections of the shore protection structure. So it is extremely dangerous to go ahead with a design unless it is a design with site specific parameters. That is what I would like to suggest.

CS smilingly remarked, I fully understand it and I wanted all others to understand. As this is an exclusive field where expertise is required, let this work be under taken by ALHW for which necessary funds would be provided by Administration after sanctioning of estimate which would be submitted by ALHW.

This is classic example of how I handled some critical situations. It was important to develop the habit of listening carefully to others. This, I feel is a balanced approach to a problem – to get all sides of the story. That way all contributors have a chance to voice their opinions and that gave multiple ways of solving a problem. I, as their leader, had a bigger responsibility to steer the team in the right direction by assimilating all the inputs, applying the best judgment and take accountability for the consequences.

SECTION 5.0

TOP LEVEL MANAGEMENT

5.0 Top Level Management

"If you pick the right people and give them the opportunity to spread their wings—and put compensation as a carrier behind it—you almost don't have to manage them."

— Jack Welch

That was the starting point for me in the untiring race of tsunami rehabilitation as my boss retired on super annuation by 31st, May, 2005 and I took over the reins of CEA of ALHW with effect from June, 2005. I was in a situation that demanded the highest order of leadership and management. Before going into the details of the historical moments of activities handled successfully under most difficult conditions, let us first take a glimpse of various characteristics expected out of a Top Level Manager.

To be a successful Top Level Manager, you should have a deep craving for making an impact, genuinely make a difference and good people skills. Certain skills of leadership can be acquired, or refined, but you should want to lead, and leading should feel like an integral part of your "being in the world."

You see the larger picture of your company, its place in the market, your place in the company, and your team's effectiveness to reach the company's goals. Despite your feelings about what your organization can do, realism must inform your decisions about issues such as: how significant is your company in the marketplace? Would anything change in the market if your company disappeared? What is your team's true ability to reach the company's goals? What is your own ability to lead your team and the company to those goals? This realism that sees the larger view may lead you to changes in your capital needs, in the configuration of your team, and in your willingness to accept new deals or change the company's direction.

You may be the person to think and remember details from various angles, various perspective and out of the box to ensure you lead your company or organization to stay ahead of all competitors. But you also must be cool headed, harmonious, highly futuristic in terms of vision, with empathy but not work on sympathy alone, ready for any challenge and willing to compete, communicate most effectively, keep the best connectedness with your employees and clients and be capable of attracting everyone to listen to you. In short, he should be a great achiever and strive to be better at it. Every. Single. Day!

I remember a story when we talk about the greatest skills and intelligence applied to achieve our goal.

One night Mr. George, an upcoming business entrepreneur was strolling in his terrace thinking of various options he wished to try in the competitive world of market business. All of a sudden, he saw some flash lights moving inside his garage where his Porche was parked. Carefully he watched and found two thieves inside his garage trying to meddle with his car. Immediately, he rang up to the police station nearby. As it was almost midnight, the police attendant who was sleeping lifted the phone with all irritation as he was disturbed of his slumber on duty. Yawning wide, he said, "Hello".

Mr. George: "I am George calling from <the address>… I am reporting on the thieves who are in my garage now. If you can send some police personnel you may nab them red handed."

Police :"(in annoyance) all police personal are on rounds. Sorry." He hung up.

Mr. George got annoyed. He stealthily came down, bolted the garage doors from outside and locked them. He went up and rang up the same police station again. "Hello, I am George calling."

"Just now I told you that nobody is here. Why are you disturbing again and again?"

Mr. George replied, "No,No.I am not calling you here now. I am only to inform you that with my gun I shot both of them dead. Whenever you have time you can come and collect their bodies. "He hung up. Within the next five minutes, a band of police personnel reached the gate of Mr. George's bungalow with their flashing red lights on their vehicles. The police person who answered the phone came to Mr. George, "Are you Mr. George?"

"Yes"

Where are the dead bodies of those thieves? You should not have taken the law in your hands!"

Mr. George opened the garage doors and the two thieves were blinking sitting in acorner of the garage. The police man saw them and looked at Mr. George.

"I thought I heard you saying that you shot them dead!"

Mr. George replied, "I thought you replied that there were no police personnel there!"

This is sheer intelligence and this is the type of intelligence required for any top level manager. In fact such intelligence groups can make every company or even any country to flourish socially and economically if they were properly treated and utilized!

With this small briefing of intelligence let us see how the qualities of a top level manager can be explained.

5.1 Harmony:

> *"Surround yourself with the best people you can find, delegate authority, and don't interfere as long as the policy you've decided upon is being carried out."*
>
> — *Ronald Reagan*

Whether you are an employee in a company, a manager who is responsible for a group of employees or the owner of a company, there are things you can do to make your workplace a better and more enjoyable place to spend the day. A harmonious workforce can be less stressful, more productive and more competitive in the marketplace.

Achieving harmony within the workplace is at the top of the list of important concerns within any industry. Creating a harmonious work environment between individuals, between management and employees, and between employers and employees presents an exciting opportunity for personal development in the exciting environment of the industries.

If employees behave in a way that disturbs harmony in the workplace, try to understand their point of view before lashing out at them. Often, conflicts at work are caused by personal issues that have nothing to do with the workplace. If you know that someone is going through divorce, illness or other difficulties, try to remember that these situations might make them difficult to deal with at times. Increase the harmony that can exist in your workplace by sympathizing with others and seeing things from their perspective.

Think about giving before you think about taking. If everyone in your workplace does this, you will almost certainly create a cooperative and harmonious team. Offers of help can range from simple things like photocopying or making coffee to undertaking major projects with a coworker. If you are on personal terms with someone else, you can also offer emotional help if she is going through a difficult time. Workplace harmony is based on the mutual belief that a team is working together rather than competing with each other. Fostering this belief by getting into the habit of offering help will create an increasingly harmonious workplace.

Interpersonal problems that disrupt harmony in the workplace often fester for weeks or months before bursting into the open in disruptive ways. Particularly if you are the manager of a workplace, work to remain aware of interpersonal dynamics amongst your team, and take steps to proactively resolve conflicts before they become serious. If two people simply get on each other's nerves, try to avoid having them work together too much. If a particular employee is bullying or disrespecting others, have a private talk with him to change his behavior before it gets out of hand.

Expanding personal communication skills, encouraging staff to work with ownership to overcome challenges, and assisting employees to identify with each other are some great ways to achieve harmony within the workplace. This atmosphere will not only help to create a better work environment. Implementing some easy strategies to create a harmonious environment will always be reflected on the bottom line.

The latest trend in bringing in business harmony is by introducing "Double drop box Technology" to the users including clients, - one for their personal folders and the other for all official and classified data. Remote wipe helps protect confidential information, account transfer helps you maintain business continuity, and sharing audit logs let you track how your Drop box for Business information is being accessed. You can have both a business account and a personal account on all your devices. The contents of each Drop box will remain separate from each other (but easily accessible).

Not many people think about giving when it comes to office harmony. It is the easiest word to say but it is also the hardest to practice. We have habituated to being selfish and to always take care of our own interests. So much so that we forget about the others who work with us. When the balance is tilt, it gives rise to disharmony.

To have harmony in the workplace give willingly. Know that it doesn't always mean someone has to lose in order for us to win. So, how do you give in order to create harmony? Very simply – remember these – **give in, give up, give out**.

Give in – do you always take a hard stance on certain issues? Learn to give in on some of these. Does it really matter one year down the road some of the stance you have taken? Yes, you need to hold on to strong work guiding principles like working with pride, passion and belief. Search yourself to see if you have any stance that is based on opinions and not principles. Learn to give in. When you learn to give in, you put in motion harmony in the workplace.

Give up – pick a personal bad habit that creates tension in the workplace. Is there something you can give up in order to create harmony? Work on it now. Today. It takes small steps in order to achieve the greater good for everybody.

Give out – here's a practice that can get you in the swing of things. You need not be rich to do this. Generously give out smiles. Smile when you meet someone in the lift.

Practice with a total stranger. A simple "Good morning", a "How are you?" or a "Which floor do you need to go?" are all practice for the simple act of giving out. Do this with colleagues too. When you do this often enough with colleagues you naturally feel good and this is contagious.

Give out a nod; a smile and a simple "How are you?" Giving does not make us any poorer; it makes us richer in our hearts. A rich heart that is giving out is a hand that is reaching out. No one can blame us for not trying.

Does creating harmony in the workplace stop here? No. Take it a step further. Be supportive of those who give too. When you see people giving, encourage them. Giving builds affinity amongst colleagues. When people can feel your generosity, they will reciprocate. Granted there will be those who will be exploitative and pounce on this opportunity. The advice is this - let them be. Only beggars take and not give.

You are rich and can afford to give. To give is your contribution to harmony in the workplace. Happiness is not a matter of intensity but of balance, order, rhythm and harmony.

"I have cherished the ideal of a democratic and free society in which all persons live together in harmony and with equal opportunities" are the words of Nelson Mandela who had personal experience in practicing what he had said.

The workplace is one of the most common places for conflicts. Considering the workplace situation, employees establish a kind of relationship among each other that keeps a diplomatic approach but usually does not go beyond personal level, though there are other relationships that develop into a deeper stage. Employees have to socialize with their coworkers because people in the workplace work collaboratively.

You may have obviously heard of the term 'office politics.' This is a workplace situation where diverse and unique personalities as well as behaviors result to an uncomfortable environment in the office. The concept of 'office politics' refers to any behavior and process that are deemed inappropriate and unfavorable for the organization. They say that in order to survive, we must learn how to 'go with the flow' or 'play the game.'

However, tolerating such condition in the workplace is not healthy for the welfare of the employees. Instead of putting up with 'office politics' or any undesirable situation, keeping a good and harmonious working relationship is a lot more advantageous. If your organization is going through a discordant atmosphere, it is high time to learn and apply the ways to maintain harmony in the workplace.

Office gossip never results to anything positive toward the organization and the individuals. It is, in fact, one of the ultimate destroyers of harmony in the workplace because it destroys trust among the people. Paying attention to gossips is no different than spreading it. This is because when you listen to gossips, it will retain in your head and affects your view of the person negatively.

If your colleagues start to engage themselves in rumors, refrain from getting involved in the tittle-tattle. Better yet, you can redirect the subject to a current work situation or other areas of interest, anything that does not highlight fellow working individuals.

Ground rules are established to keep an organization guided and controlled. At times, we feel that certain rules are choking us to the neck. This may be grounds for emerging conflicts between employees and the management. Whenever an individual feels that there are inconsistencies and irrationalities in the workplace, tendency is the person struggles to break free from such process.

For employees, it may take an effort to really understand that setting such rules is a way of maintaining standards in the company. But this should be well-implemented and properly disseminated by the management in a way that will not seem like a threat to the individuals.

"Always aim at complete harmony of thought and word and deed. Always aim at purifying your thoughts and everything will be well" are the words of Mahatma Gandhi who strived hard to achieve harmony in the freedom struggle.

On a rainy night of New Year's Eve, a lonely little girl was saddened by the disagreements that brought upon her family members.

Little girl sadly said, "It's Chinese New Year's Eve, but it's been raining all day long. New Year is coming but everyone is not in a good mood. Brother insists on buying a new motorcycle. Dad, Mom, and Brother are quarrelling about it again."

Dad told to Brother, "Can't you use the old motorcycle? You just know how to spend money!"

Mom also said to Brother, "The economy is bad, why don't you just use the old one?"

Brother angrily said, "The old motorcycle breaks down all the time, you can use it!

I won't!"

"It's your fault. You have spoiled him!" Dad said to Mom.

Little girl asked herself, "It's New Year, but why is everyone unhappy?"

"Kitty, can you tell me how to make everyone happy for the New Year?" asked the little girl to her cat.

Then, outside the window, the little girl looked four old men were coming, "Dad, Mom, look! Brother, come and see, quick!"

Mom said to the old men, "Ah, it's raining so heavily, Sirs, please come in for shelter!"

Old man #1 said, "Ha-ha, thank you for your kindness, Madam. But we have a rule, only one of the four of us can come in. Who do you wish to invite in?"

The four old men then introduced themselves one by one.

Old man #2: "I am Wealth."

Old man #3: "I am Success."

Old man #4: "I am Well-being."

Old man #1 said with a laugh, "Ha-ha, everyone call me Harmony."

Dad said to the four old men, "Surely we should invite in Wealth, then we can have a comfortable life!"

Brother said, "No, no, choose Success! I want my family to be proud of me!"

Mom said, "Wait a moment! I think Well-being is the most important!"

Dad exclaimed, "Wealth!"

Brother exclaimed, "Success!"

Mom exclaimed, "Well-being!"

Little girl asked to her mother, "Mom, Mom, What is harmony? Why don't you invite in Harmony?"

Dad said, "Yes, you're right! Why don't we invite in Harmony? New Year is here, we should be harmonious. Let's invite Mr. Harmony in, then!"

Seeing that all of the four old men came in together, little girl's father said, "Eh? I thought you said only one of you can come in? Why did all of you come in?"

Old men #1 replied, "Ha-ha, we have another rule. If Harmony enters, Well-being, Success and Wealth will follow."

Little girl happily said, "Now I understand. To be happy is to be in harmony."

This is a Chinese story by Ven. Master Xin Yun (Master Hsing Yun) which says when Harmony is established, everything falls in place.

5.2 Futuristic:

> *"Management by objectives works if you first think through your objectives. Ninety percent of the time you haven't."*
>
> **— Peter Drucker**

The Managers at the Top Level should have a clear vision of the goals and policies of their business and there should be no hesitation in upgrading those as the business in the world changes.

Imagine that there is a cage containing five monkeys. Inside the cage, hang a banana on a string and place a set of stairs under it. Before long, a monkey will go to the stairs and start to climb towards the banana. As soon as he touches the stairs, spray all of the other monkeys with cold water. After a while, another monkey makes an attempt with the same result - all the other monkeys are sprayed with cold water. Pretty soon, when another monkey tries to climb the stairs, the other monkeys will try to prevent it.

Now, put away the cold water. Remove one monkey from the cage and replace it with a new one. The new monkey sees the banana and wants to climb the stairs. To his surprise and horror, all of the other monkeys attack him. After another attempt and attack, he knows that if he tries to climb the stairs, he will be assaulted.

Next, remove another of the original five monkeys and replace it with a new one. The newcomer goes to the stairs and is attacked. The previous newcomer takes part in the punishment with enthusiasm! Likewise, replace a third original monkey with a new one, then a fourth, then the fifth. Every time the newest monkey takes to the stairs, he is attacked. Most of the monkeys that are beating him have no idea why they were not permitted to climb the stairs or why they are participating in the beating of the newest monkey.

After replacing all the original monkeys, none of the remaining monkeys have ever been sprayed with cold water. Nevertheless, no monkey ever again approaches the stairs to try for the banana.

Why not?

Because as far as they know that's the way it's always been done around here. This is not the way a business top brass thinks. He has to break all shackles and think differently.

If you're assuming leadership of a large organization or department, take the time to understand its current trajectory. Making too drastic and immediate a change can derail both confidence and long-term strategy.

Setting goals that challenge everyone in the organization to strive for better performance is one of the key aspects of the futuristic planning process. Goals must be aggressive, but realistic. Organizations cannot allow themselves to become too satisfied with how they are currently doing--or they are likely to lose ground to competitors. The goal setting process can be a wake-up call for managers that have become complacent. The other benefit of goal setting comes when forecast results are compared to actual results. Organizations analyze significant variances from forecast and take action to remedy situations where revenues were lower than plan or expenses higher.

Management consultant Peter Drucker had a great saying concerning futuristic thinking. He said, "**The best way to predict the future is to create it**".

There are literally thousands of answers to the question. Futuristic thinking involves choosing the answers which best shape the future you want to create. If you don't like something change it, if you can't change it, you might consider changing yourself, if you can't change yourself to that degree, you need more futuristic thinking rather than giving up on the strategic issue.

You must not be totally dependent on another person (key employee or partner) to do all the heavy lifting in your company for generating the capital and revenue that is its lifeblood. It weakens your position in the company and your control of your future. Even if that partner is loyal and committed to you and the company, someday the bus runs us over when we are not looking. Loss of that person for any reason can cost you your company if you cannot carry your own weight in closing.

In our fast-moving world of sound bites, information overwhelm, time-shifting and 24/7 availability, it is rare for a CEO to hear any advice that includes having patience. But you must have it, and the capital to sustain it, or your company will fail.

You don't have to build your entire vision all at once, or at the outset of your new venture. If you have the patience to plan and execute carefully, while still being agile, your patience will pay off in those hidden strategic assets that make a company successful (clarity, responsibility through the ranks, market responsiveness, and a single collective vision of the company's goals).

Successful futuristic thinkers usually follow a process something like this:

* Decide what they want
* Spend time thinking about why it's important to themselves or someone else
* They take a small step toward making it a reality
* Step back and think about what they learned on that small step
* Continue lots of thinking integrating what they learned into developing a next step

Today the term futurist applies to visionary leaders, innovators, thinkers, writers, consultants, presenters and others who "look to the future" and just as frequently to those who "provide analysis of the future" via such methods as visioning, intuition, analogy, argument, logic, planning, policy analysis, cultural criticism, strategy development, marketing, road mapping, goal setting, forecasting, modeling, statistics, trend analysis, operations research, investment, surveys, horizon scanning, scenario development, prediction, prediction analysis, prediction market development, risk analysis and management, and other future-oriented activities.

You make judgments about opportunities, people, their actions, your actions. Some of the first judgments will be about control. Do you share control with your founding teammates, because they are sharing your risk?

What if they are sharing some of the risk, but none of the capital risk? Is it equal then? Do you use stock ownership as the basis for control, or do you isolate control for yourself (if you are taking the majority of risk) or share it with a small executive board? And when new owners and capital investors arrive, how does that control shift? These decisions must be made early on, with an eye to the larger view mentioned above.

You make the tough decisions, the ones that make others join up with you, and even the ones that cause you or others grief (personally or professionally). For example, you may need to remove a close friend from your startup team during year 2, after extensive commitment from the friend, when that person no longer is a significant contributor to the future of the company. Or, you must choose to accept or reject an influx of needed capital based on deal terms that may change the direction of your company.

You accept a certain unique isolation, because the buck stops with you, because very few around you can understand the pressures you experience every day, and because to lead is to carry the burden of these responsibilities, consequences and isolation. I have often worked with CEOs who began our initial conversations with, *"I don't have anybody who can understand what I need to do and help me make the decisions I need to make."*

Planning is always done for future and future is uncertain. With the help of planning possible changes in future are anticipated and various activities are planned in the best possible way. In this way, the risk of future uncertainties can be minimized.

For example, in order to fix a sales target a survey can be undertaken to find out the number of new companies likely to enter the market. By keeping these facts in mind and planning the future activities, the possible difficulties can be avoided.

While we should live knowing that tomorrow may never come, we also know that it is wise to plan for the future. *Proverbs 6:6-8 says we should take a lesson from the ant that plans for the future. Proverbs 16:9 says that we should plan and allow God to direct in and through our planning. If you don't sit down and count the cost of a project, there is no reasonable way to expect that it will be accomplished. Planning is necessary. While you must plan, you should also be flexible in those plans and allow the Holy Spirit to guide you (Proverbs 16:9; 2 Corinthians 5:7).*

Leadership is exciting and dynamic and stressful and not for everyone. Understanding what you must do to lead a new company as its CEO is critical to its success, and yours.

5.3 Empathy:

> *"Don't equate activity with efficiency. You are paying your key people to see the big picture. Don't let them get bogged down in a lot of meaningless meetings and paper shuffling. Announce a Friday afternoon off once in a while. Cancel a Monday morning meeting or two. Tell the cast of characters you'd like them to spend the amount of time normally spent preparing for attending the meeting at their desks, simply thinking about an original idea."*

> *— Harvey Mackay*

The best executives "have an affinity for others, they're affable, they like people. They excel at maintaining the self-esteem of others," says Benton. Just ask Jack Mitchell, CEO of luxury retailers Mitchells/Richards/Marshs. He is the kind of warmhearted boss everyone would like to have. In his book, "Hug Your People," he promotes being nice to employees, trusting them, recognizing them and of course, hugging when appropriate.

Empathy is about standing in someone else's shoes, feeling with his or her heart, seeing with his or her eyes. Not only is empathy hard to outsource and automate, but it makes the world a better place. says Daniel H. Pink

Compassion is a human emotion prompted by the pain of others. More vigorous than empathy, the feeling commonly gives rise to an active desire to alleviate another's suffering. It is often, though not inevitably, the key component in what manifests in the social context as altruism.

In 2008, Karina Encarnacion, an eight year-old girl from Missouri, wrote to President-elect Barack Obama with some advice about what kind of dog he should get for his daughters. She also suggested that he enforce recycling and ban unnecessary wars. Obama wrote to thank her, and offered some advice of his own: "If you don't already know what it means, I want you to look up the word 'empathy' in the dictionary. I believe we don't have enough empathy in our world today, and it is up to your generation to change that."

This wasn't the first time Obama had spoken up for empathy. Two years earlier, in a commencement address at Xavier University, he discussed the importance of being able "to see the world through the eyes of those who are different from us—the child who's hungry, the steelworker who's been laid off, the family who lost the entire life they built together when the storm came to town." He went on, "When you think like this—when you choose to broaden your ambit of concern and empathize with the plight of others, whether they are close friends or distant strangers—it becomes harder not to act, harder not to help."

An example of empathy filled with sympathy is:

Sam walks into his boss's office and says "Sir, I'll be straight with you; I know the economy isn't great, but I have over three companies after me, and I would like to respectfully ask for a raise."

After a few minutes of haggling the boss finally agrees to a 5% raise, and Sam happily gets up to leave. "By the way," asks the boss, "Which three companies are after you?"

"The electric company, water company, and phone company!"

In helping others, we can often get a better view of what we're capable of. It will also help you feel better about yourself. Helping others brings a wonderful sense of fulfillment and you will find yourself more confident than ever.

A doer with such an overwhelming vision builds the pieces that can be adopted by the market, watches the changes brought about by that adoption, then builds the next and the next, in sequence, as the market is ready for each next step. And many of the next steps are adjacent to his original vision, or a new technology arrives that changes the vision itself.

5.4 Connectedness:

"Management is efficiency in climbing the ladder
of success; leadership determines whether the ladder is
leaning against the right wall."

— *Stephen R. Covey*

Research has shown that two benefits of positive workplace connections are increased productivity and low turnover. Not surprisingly, the same research shows a correlation between the lack of connectedness and disengagement. With this in mind, it is important to develop a work environment that fosters the building of these connections. Once this environment has been developed, business owners and management also need to be pro-active in encouraging and providing the opportunity for these connects to be made in the workplace, with an awareness towards ensuring that these connections are kept positive and do not turn into the cliquish type behaviors typical of human beings. If this is seen, steps should immediately be taken to right the ship in this area, to make sure employees do not feel isolated from the group. Employers need to also have a pro-active conflict resolution system in place to quickly address negative behaviors and interactions. Negative behaviors such as gossiping, tattle-telling, and backstabbing should be strongly discouraged.

Here are some ways to build connections in the workplace:

Organize informal get-togethers away from the office; this could be a baby-shower, a birthday celebration or an unplanned coffee break & walk around the block

One great way for employees to interact is sharing food; plan a luncheon to celebrate a company success, the completion of a project or the winning of a new account/customer

Business owners & managers should take the time to let employees know they are interested in them as a person. Ask questions to learn more about them, such as what they did over the weekend or what their favorite vacation has been.

Have empowering staff meetings and regularly engage in team building exercises

Richard and Angela Gent, founders of the Connectedness Research Group, have created a way to measure the impact of employee connectedness. Their data shows that interventions that increase connectedness at work can result in fewer sick leaves, more employees promoting their organizations as a great place to work, increased loyalty and longevity in the job and a 20% increase in personal well-being. With these benefits in mind, positive connections should be a focus area and goal for any organization.

More often than not, our work is that of co-creation. People still need and want to come together in a physical space. They come for very simple reasons: people need people, people need technology and people need spaces that bring those two together in effective ways. We also need ties and trust to those we innovate and create with. Coming together in a shared space is still one of the best ways to build these ties. For all these reasons, a workspace still matters very much to small businesses and small businesses are leading the way in creating the post-cubicle world.

Feelings and emotions can affect our work in ways we don't even realize. High-performing leaders understand that staying positive – not holding grudges, dwelling on problems, or spending time focusing on negative influences -- is an integral part setting and accomplishing their goals.

Collaborative, flexible and customizable spaces encourage communication and connectivity. Designs that also include suggested zones for informal meetings promote impromptu team based gatherings and a productive environment.

A strong leader is a great communicator – someone who's both an inquisitive conversationalist and an engaged, active listener. Everyone wants to be heard, and etiquette coach and entrepreneur Jacqueline Whitmore says you can start connecting better by asking more open-ended questions, such as "Tell me, what did you enjoy the most about the conference?"

While every organization in the world is unique, most have characteristics in common. Each has its own culture that derives from its mission, vision, values, and the people who work there. Every employee has his or her own roles, responsibilities and work style, but he or she seldom works in isolation, even if working individually. All employees are ultimately connected to one another through a common purpose to succeed and fulfill the mission of the organization. This is the nature of healthy organizations.

While every organization in the world is unique, most have characteristics in common. Each has its own culture that derives from its mission, vision, values, and the people who work there. Every employee has his or her own roles, responsibilities and work style, but he or she seldom works in isolation, even if working individually. All employees are ultimately connected to one another through a common purpose to succeed and fulfill the mission of the organization. This is the nature of healthy organizations.

High performing individuals commonly enjoy even higher levels of connectedness within the organization. Through experience, they've developed their own networks consisting of connections that span across organizational bounds. Such networks are characterized by influence, not authority. Coworkers are united in the interest of organizational success.

Today, social media connect people around the world. Those of us in the workplace can engage across these media to establish professional relationships with others with similar expertise or professional interests. While building these professional networks, we still protect confidential information as we have for years, but a great many aspects of work can be enhanced by interacting with others: new ways of approaching problems, discussing abstract concepts, and shared challenges with common tools are just three examples.

Globally connected workers, those individuals who build and maintain effective networks both inside and outside their organizations, have the potential of achieving the highest levels of connectedness. Only these self-leaders can successfully bridge the barriers that separate the workplace from the public networks. Anyone can search the web for information, but globally connected workers can bring the collective wisdom of thousands to bear, when it's needed most, to better their work product and bring more value to their organization.

Having a strategy to create Employee Connectedness amongst new and existing employees is crucial to maximizing employee willingness to deliver great customer service. Stanford University recently released a study that showed 87% of success in business is based on connecting with people and only 13% on product knowledge. Customer service staff pass on to their customers their experience as an employee. If staff is unhappy you sense this as a shopper and it erodes your experience. If customer service staff experience low Employee Connectedness, they will pass this on to the customers they should be trying to connect with.

A United Nations report released during November, 2011 has for the first time established a Connectedness Index that measures countries' knowledge networks and links their level of connectedness with economic development indicators.

To determine the Connectedness Index, the report takes into account international, inter-organizational and intra-organizational networks established by each country.

The report provides a global connectedness ranking, which was topped by Switzerland and followed by Sweden, The Netherlands, the United States and Finland.

The report, jointly produced by the United Nations Industrial Development Organization (UNIDO) and the Centre for Global Governance Studies in Leuven, Belgium, argues that a country's connectedness has a strong positive impact on its economic and industrial development.

"Economic success, social cohesion and environmental sustainability of a country depend more than ever on the performance and behaviour of its neighbours, regional leaders and global economic powers, making knowledge management and knowledge networking crucial for policymakers," said UNIDO in a news release.

This makes it clear how important the connectedness is in any organization or in any government.

5.5 Communication:

> *"Good management is the art of making problems*
> *so interesting and their solutions so constructive that*
> *everyone wants to get to work and deal with them."*
>
> —*Paul Hawken, Natural Capitalism*

Your communication should not be left "to be understood". Be sure that you are clearly communicating that what you want to convey is definitely "not misunderstood." Remember, Communication is not a one way street!

When you want to communicate something to somebody, make sure that you tell them the subject matter you want to talk about, the time you may take to finish and what exactly you want them to do after you talk to them. Do not take anyone granted that they don't have any business to do except listening to you. Be precise, short and sweet.

Always keep one thing in mind when you call a client. That is, you should never take it as granted that they are free at any time you call them. Please ensure that you appreciate the value of their precious time and do mention to the effect "Sir, I realize that your time is very precious and I won't take much of your time." This will relax them and make them ready to listen to you. You should be always alert to gauge their response and attention while you talk to them and only and only if they show interest and wish to continue talking you are a successful seller of your product.

Similar approach is essential when you plan a meeting with your boss in your organization. He should feel comfortable with your timing, duration of the proposed meeting and also your briefing to him before the meeting so that he is well prepared. Otherwise, it becomes your big problem to wipe out the opinion he forms on you regarding your communication skill as well as organizing capacity. It is always to get the time fixed as per their convenience.

Your strategic course of action is only as effective as your ability to communicate it. Have the pipeline and protocol set up to get your message out there, and don't forget that communication goes both ways.

When things are happening around you and you have an opinion or know a better way to do something, speak up! Don't just let your life happen around you. Take active part and be in control of the situation. This shows you that you are capable of being in control and leading others.

Speech has powerful potential: it can lead to inspiration, friendship and growth – yet it can also lead to destruction and harm. Before you speak – check your purpose and motivation. Consider asking yourself these three questions: What I am about to say, is it true? Will it lead to the benefit of others and harmony? Is the recipient ready to hear this? Communicate in a way that promotes harmony and good, while avoiding any gossip or talking maliciously about others.

Make sure your initial communication is clear. The first barrier to communication is unclear transmission of a message. Unless we are trained in clear, unambiguous and effective communications, we often send unclear messages to others. Many people have a "shorthand" form of communication that may be understood by intimates but is virtually incomprehensible in the office environment. Clearly state what result you want to obtain and, if necessary, the steps to obtain it.

Make sure that you've been heard. We all filter what is being said by another person according to our expectations, culture, upbringing and experience. Even if one person is incredibly clear in his or her communications, it is unlikely that others will hear what is actually being said.

Today's corporate environment is enough to make Gandhi need a Valium. The last thing taken into account in corporate decisions is the human factor. Keep in mind that today's stress-filled, "do more with less" office environment is an accident waiting to happen. Clear communication, a little compassion and taking a few extra moments to make sure everyone is on the same page will go a long way toward a productive office environment.

Effective managers and supervisors already know a few things about communicating with employees. Communication helps to build relationships, promotes mutual understanding, and enables employees to contribute to organizational success. Moreover, it's a skill that can be learnt – but it takes practice.

Achieving success through effective business communication typically involves acquiring skills and experience. Learn to express your ideas clearly and succinctly. Take steps to understand your audience. Target your message to meet their needs. Using effective communication helps your message get interpreted correctly. Avoid conflicts with co-workers, managers and customers by using effective communication techniques. Practice using techniques, such as active listening and paraphrasing to ensure your success.

Keep track of the questions people ask you and learn how to respond to common inquiries. Use active listening techniques, such as paraphrasing what you have heard or nodding in acknowledgment. Ask clarifying questions yourself to ensure you truly understand what your audience does not comprehend. Ask open-ended questions to start a conversation, get more details or get input on issues. Ask closed questions that require a simple "yes" or "no" answer to confirm your understanding, get agreement or conclude a meeting.

Choose the right communication format for each situation. For example, avoid using email to communicate emotional issues, such as bad news. Use written communication to convey lists of information, such as policies and procedures. Use diagrams and charts to summarize complicated financial data. It is also important to know what mistakes are to be avoided during communicating.

Stever Robbins is president of VentureCoach.com, a Cambridge, Mass. entrepreneurial coaching service. He talks about seven communication mistakes the managers are likely to make and cautions against the same.

1. **Making controversial announcements without doing groundwork first**

Any controversial decision can engender rumors, anxiety, and resistance. So rather than announcing a controversial decision to an entire group, prep people one-on-one. Learn who will object, and why.

Decisions about change are the most charged — reorganizations, changing goals, and the departure of key employees create uncertainty, and uncertainty generates anxiety.

To forestall anxiety, open a dialogue with the other person. Put a name to the problem: "This reorganization means we'll be doing some things differently, and that makes some people apprehensive." Then address the concerns raised in response to your statement:

- Is the other person uncertain about the future? Share the scenario you expect to unfold.
- Does the reorganization jeopardize a project? Share plans for keeping it afloat.
- Demonstrate that you get it, keeping in mind that you can address emotion better with body language than with words. Make sure yours conveys concern and empathy.

2. **Lying**

Some lies or partial truths are well-intentioned. Certain topics must remain confidential while they're under discussion. But be careful how you keep secrets. If people know you've lied, you will lose their trust forever. A start-up company's controller watched the CFO lie to members of other departments and subsequently began to doubt the CFO's sincerity. He began looking for a new job with a boss whose intentions he could trust. In that instance, lying cost the company a valuable employee.

Rather than lie, train yourself to respond, "I'm not free to comment" or "I can't answer that fully right now," when asked about confidential or sensitive topics. Consistency is important. Warren Buffett never discusses his investments, even with shareholders. As a result, his silence on a particular deal gives away nothing.

3. **Ignoring the realities of power**

Surprised that you never hear bad news until it's too late? Don't be. The more power you have, the less you'll hear about problems. It's human nature: problems are filtered and softened as they ascend the corporate hierarchy, with each messenger seeking to soften the blow. If you want an honest assessment of a problem, seek out bad news. Welcome it. And when it comes, show your appreciation.

Conversely, messages are magnified as they travel down the hierarchy. If you look pained during a presentation, everyone will "know" you hated the presentation (or worse — the presenter). No one will think to blame the pastrami sandwich you ate too fast before you came to the meeting. Jokes are especially dangerous. When the managing director of a consulting firm joked, "If you're not here Sunday, don't bother coming in Monday," his project team wasn't sure what to do. One said, "We were pretty sure he was joking, but. . ."

Put a lid on rumors by using plain, simple language. End meetings by reviewing your reactions and next steps. "I value your analysis, Chris. The sales trend is disturbing. Let's follow up on Wednesday."

4. **Underestimating your audience's intelligence**

It's tempting to gloss over issues because "people won't understand." Why explain a reorganization when you can simply say, "Here's the new org chart"? But that's a cop-out. Front-line employees may not be masters of organizational design, but they deserve to know the rationale behind changes that affect their lives. If you think your people won't understand something, remember it's your job to explain it to them.

Many managers like to gloss over problems when motivating their teams. But if things aren't going well, those teams are probably well aware of the problems. In fact, they've probably known about them longer than you have. Rather than avoiding the situation, enlist their skills in finding solutions.

5. **Confusing process with outcome**

In goal-setting, compensation, and evaluation, it's easy to confuse process with outcome. You promise your team a 7% raise, but then the board, concerned about the downturn, caps raises at 3%. You fight like mad to raise the number, and you compromise on 4%. But your people don't appreciate it. In fact, they're downright resentful. How could they be so insensitive to all your hard work?

Simple. Your hard work was process, and you promised them an outcome. You want them to appreciate how hard you tried, but they wanted a specific result. Since they didn't get it, they can't see past that fact. You want people to value you for your hard work. But when evaluating others, it's always easier to judge outcomes. Most organizations penalize employees for the wrong outcome, even if they follow the right process. Perversely, others are rewarded for the right outcome, even when they flout the rules about process.

6. **Using inappropriate forms of communication**

E-mail is great for conveying information, but don't use it for emotional issues; e-mail messages are too easy to misconstrue. If you're squirming while reading an e-mail, leave your computer and deal with the situation in person or by telephone.

At the same time, phone calls and face-to-face meetings are inefficient ways to disseminate information, but great for discussing nuanced issues. You can respond directly to the listener's reaction, and you can use your tone of voice and facial expressions to control your message. "I'm sure you did a great job" could be read sarcastically in an e-mail, but the same words can be delivered sincerely in person with the right tone of voice.

Furthermore, some people are listeners, while others are readers. Listeners won't focus on written memos but are great in conversation. Readers write great memos and are also glad to read them, but conversation sometimes fails to fully engage them. If you talk to a reader or write to a listener, your message might not get through. Don't be afraid to ask people how they prefer to receive information; most people know the answer. If they don't, a little attention on your part will reveal what works best. (And for some people, it's a combination of the two.)

7. **Ignoring acts of omission**

What you don't say may be sending as loud a message as what you do say. If you don't give praise, people get the message they're unappreciated. If you don't explain the rationale behind decisions, the message is that you don't trust them. And if you don't tell people where the company wants to go, they don't know how to help it get there.

When fundraising became the CEO's priority at a distance learning company, he stopped communicating his vision to employees. Since money was constantly on his mind, he did mention financial goals. Eventually, the company culture became money-focused, and the vision was lost. But when the CEO delivered a vision-oriented presentation at a conference, one of his employees approached him afterward to say that she had never felt so inspired. As a result, he changed his internal communication strategy to emphasize vision once more, and saw morale soar.

By their very nature, mistakes of omission are hard to uncover. Review your major goals and the communication that's needed to support those goals. Ask what message may have been sent by your silence so far. And be willing to ask people, "What messages are you getting from me?"

That makes things clear as to what to do and what not to do while communicating for effective management.

5.6 Competition:

> *"Focus on a few key objectives... I only have three things to do. I have to choose the right people, allocate the right number of dollars, and transmit ideas from one division to another with the speed of light. So I'm*

really in the business of being the gatekeeper and the transmitter of ideas."

— Jack Welch

In this highly competitive world of business, it is essential to have up-to-date knowledge of everything going on around in the field of business. The scientific and real data analysis makes it possible to a good extent for which data collection is most important.

"Give me a little data and I'll tell you a little. Give me a lot of data and I'll save the world" says Darrell Smith, Director of Facilities and Energy Microsoft which tells the importance of data without which you are lost in the competition.

Planning helps organizations get a realistic view of their current strengths and weaknesses relative to major competitors. The management team sees areas where competitors may be vulnerable and then crafts marketing strategies to take advantage of these weaknesses. Observing competitors' actions can also help organizations identify opportunities they may have overlooked, such as emerging international markets or opportunities to market products to completely different customer groups.

If we only take the easy route, it can be easy to think that we aren't capable of doing things that are hard. Prove to yourself that you can take on challenges by doing just that: take on challenges. Do things that will be rewarding, even though they'll be hard work. You can do it!

Successful leaders recognize the need to adapt to the ever-rapidly changing ways to do business in the global environment. These leaders seek to build competitive advantages around the core competencies of the organization while also reducing costs to conduct their business.

These organizations also understand that doing the best that they do is not always enough to be on top. In order to keep the competitive position in the domestic market, they will need to acquire knowledge of other key competitors in the global marketplace. They need to stay informed of other domestic and foreign competitors' potential strategies, as well as their competitor's strengths and weaknesses. Global competitors understand that with increased competition, new ways to differentiate their products and services need to be developed. One of the best motivations to innovate for a company knows that they can lose business to the opposing team.

The competition in business leads to stress. Stress Management is an important aspect every manager should be aware of and practicing. *Stress Management is jovially explained as below:*

"Picture yourself near a stream.
Birds are singing in crisp, cool mountain air.
Nothing can bother you here.
No one knows this secret place.
Hear the sounds, and enjoy the peace.
Feel the sweet air on your face, your skin.
You are in total seclusion from that place called the world.
The soothing sound of a gentle waterfall fills the air with a cascade of serenity.
The water is clear.
You can easily make out the face of the person whose head you're holding under the water.
There now... feeling better?"

A clear understanding of your competition is key to the success of any business. Even if your product or service fills a unique gap in the market, there are always other companies offering something similar, or there are other ways to satisfy the same customer's need. The key when thinking about your competition is to learn what makes the customer choose one product or service over another. The different options that customers consider are usually competitors.

More competitions will definitely boost the industry and drive economics. So, for those who run a business or those just starting out on your own, you should know who are your "DIRECT" and "INDIRECT" competitors.

Let's talk on direct competitors. These can be classified as

1) Offering the same products and/or services as you are offering to your clients and/or customers.

2) Having the same targeted field of clients, customers and/or demographics.

3) Using the same tactics in advertising or bringing news/information of products/services to your targeted demographics.

A very good example of a DIRECT COMPETITOR will be GOOGLE and YAHOO.

On to Indirect competitors, these competitors are a bit hard to spot as compared to direct competitors. They can be classified as "wanting to have a share of the pie".

Motivated by the desire for achievement, Achievers have goal-oriented lifestyles and a deep commitment to career and family. Their social lives reflect this focus and are structured around family, their place of worship, and work. Achievers live conventional lives, are politically conservative, and respect authority and the status quo. They value consensus, predictability, and stability over risk, intimacy, and self-discovery.

Successful people use creative strategies to reach their goals. They look at options and make informed decisions. Successful planning requires that you know your rights and responsibilities and strengths and challenges; set goals; work toward those goals; and use tools and resources available to you. One key skill for success is self-advocacy. Being able to self-advocate requires that a young person become an expert on their disability, know what specific services and help they need, and be able to use strategies to obtain this help and support. One's life should not be defined by the assumptions of others.

Knowing and valuing yourself, setting goals, and planning help build important foundations, but action is required to make your dreams come true. To take control of your life it is necessary to choose and take appropriate action. Take charge. Move forward.

Strategic management involves the formulation and implementation of the major goals and initiatives taken by a company's top management on behalf of owners, based on consideration of resources and an assessment of the internal and external environments in which the organization competes.

Strategic management provides overall direction to the enterprise and involves specifying the organization's objectives, developing policies and plans designed to achieve these objectives, and then allocating resources to implement the plans. Academics and practicing managers have developed numerous models and frameworks to assist in strategic decision making in the context of complex environments and competitive dynamics. Strategic management is not static in nature; the models often include a feedback loop to monitor execution and inform the next round of planning.

5.7 My journey as Top Level Manager:

> *"I believe the real difference between success and failure in a corporation can be very often traced to the question of how well the organization brings out the great energies and talents of its people."*
>
> — *Thomas J. Watson, Jr.*

Having seen the predominant characteristics of a top level Manager, let me relate my experiences on my assuming the charge of CEA after the retirement of my boss by May, 2005.

That was the time I was put on the track to take a sprint on the rehabilitation of damaged port and harbor structures in Andaman and Nicobar Islands where shipping is the lifeline.

That was the time when the quality of a Top Leadership was demanding to be displayed with unassuming certainty in a telling manner.

Firstly, it was my responsibility to create a perfect "Harmony" among various sections, i.e., my team of Engineers and supporting staff, the port users, the Andaman and Nicobar Administration, Union Ministry of Shipping, and the chain including Planning commission. I was determined to stand up to the occasion as was confident in my technical and managerial skills.

Of course my team of Engineers and supporting staff were very much aware of my working style and confident of my leadership as they have been working with me on several critical projects. As far as the A&N

Administration was concerned, they have seen my performance in setting right the Mus Harbor in Car Nicobar as explained earlier apart from various other projects. My main strength at that point of time was Mr. Naresh Kumar, a senior officer from Indian Administrative Service, who held the charge of Secretary (Shipping) in Andaman and Nicobar Administration. He took keen interest in the rehabilitation of our islands. Almost every day, we used to take the helicopter in the morning itself to each of the affected islands from Port Blair, conduct a thorough study on case to case basis, decide actions right on the spot and gave clear directions to staff to proceed with the rehabilitation efforts.

All the technical details were taken care of by me while all logistical support was given to me by him whole heartedly in the interest of islands restoration. We both have briefed the Union Ministry of Shipping as well as Planning Commission in periodic status meetings. We worked together and fixed priorities of various rehabilitation activities based on the demand and priority as the list was running more than a hundred items.

We categorized the list of projects as Short Term, Medium Term and Long Term projects so that we were clear of what was to be done and when it was to be done. This list was discussed in detail in the Union Ministry of Shipping, and subsequently with Planning Commission of India and got approved by all concerned.

During such discussions, there were high tension voltages too, flashing on certain issues. For example, when we were discussing on rehabilitating the damaged break water of Hut Bay, one of the officers present was strongly recommending that the rehabilitation should be done in such a way that the break water should be capable of withstanding tsunami waves and such design should be adopted while rehabilitation was done.

Before going further, let me explain in simple terms what break water is. When a harbor is directly exposed to open sea without any protection from high waves which may attack during monsoons, the maneuvering of ships from and to the harbor becomes difficult. Hence it becomes essential that certain engineering measures are to be considered to arrest and break the fury of high waves to provide a reasonably calm harbor basin. A structure constructed at the entrance of the harbor in order to break such wave actions and to provide tranquil harbor basin inside harbor is called break water.

In the meeting, the officer kept on insisting for redesigning the break water to be rehabilitated so that it can withstand any such tsunami in future and everyone seemed to be inclined to agree with him. Slowly, I started speaking," Sir, your contention for stronger structure is well taken. You were mentioning that it should be designed to with stand tsunami waves. Can you please explain as to which intensity of tsunami you are referring to?"

"Why? What we have just experienced one now!"

"Sir, We do not know when such tsunami is going to hit again. It could be tomorrow or after 25 years or after 100 years too. Moreover, we don't know that the next tsunami when it happens would be of same intensity as we experienced now. We have to keep in mind that the technology is also changing day by day and who knows! There would be better and effective design available in next few years. Over designing means unacceptable increase in project cost and timelines too. So, I personally feel that we should not go for a higher design than normal as in any case, we can redesign if required with new technology whenever it warrants."

This was a clear logic no one could deny and finally it was agreed upon to design for normalcy.

With the able support extended by Mr. Naresh Kumar, we could complete all the short term rehabilitation projects in six months and also progressed well in other areas too. The Long Term projects consisted mostly of such projects which were to be totally reconstructed or fresh projects which were not existed before tsunami.

During one of our trips of tsunami rehabilitation, we hopped down to Katchal Island. This island was also largely damaged by the tsunami waves. The harbor structure which existed earlier, the jetty, got completely damaged and had disappeared totally. This was the jetty being used as a place of recreation by the local population in the evenings. As there was no other entertainment facilities in these remote islands, people prefer to gather in jetty premises in the evenings, sit, chat, gossip, share tea, *pan* (flavoured betel leaves with betel nuts and spitting in all available corners giving a red paint feel).

On the day of 26th December, 2004 which was a Sunday, many of the school teachers who contribute to the population of Katchal had gathered in the beach area after the dreadful earthquake which rocked the islands in the wee hours. By then, they watched the sea receding unusually inside thereby exposing more area of the beach even below lowest low tide due to which the fishes in the area got stranded on exposed land and were fluttering on the sea bed! This was a scene never seen earlier and so many of them started running towards the sea to collect as much fish as possible free of cost. Alas! They never knew that it was their death call from nature. When they were busy collecting fish, there came the first tsunami wave, rising to a height of about 6 to 8 meters like a vertical wall of water with the top portion of the snake head. What a colossal damage! Thousands of them were washed away leaving no trace. Later some bodies were recovered from deep inside the green forests, some hanging on trees and many bodies totally missing. Due to subsidence of the land mass after the major earth quake, the ingress of the sea was more and more area of land was swallowed by the sea. The people at Katchal were stranded as there was no harbor structure for ships to come, which rendered a horrendous situation that, no food or ration items could reach Katchal. It had become our top most priority to develop some sort of a temporary berthing structure at Katchal to receive rehabilitation materials and ration items for the people of Katchal.

5.7.1 Rising to the Occasion:

> *"Management is, above all, a practice where art, science, and craft meet."*
>
> *— Henry Mintzberg*

As urgent rehabilitation of the damages were to be done there as explained earlier, it was extremely essential to construct some temporary arrangement for berthing of at least smaller ships of 3 meter draft. We had done extensive search along the Coast of Katchal Island and found that the location in Kapanga village of Katchal where a small rivulet joins into the sea could be the place where a temporary jetty could be constructed to facilitate rehabilitation. We immediately jumped into action as there was no other feasible solution. We located a few old steel containers of smaller capacity which would have been used to store manures. Its sideways dimensions were about 3 meters long and 3 meters wide. We moved 3 such containers to this selected location at Kapanga village, fixed them in a row alongside the bank of the river flow at a suitable location to form a temporary berthing face for smaller ships to approach. We made it secure and strong by filling it with all concrete debris strewn all around, concreted the top 20 cm thick with cement concrete which could serve as landing area. We added a few number of old truck tyres to the front face of this temporary jetty so they will act as fenders to take the berthing load of smaller ships and to provide safe cushion while berthing of small crafts. This temporary jetty was commissioned by 15th, March, 2005, just in a little more than two months after tsunami, which was extremely helpful in accelerating Katchal's immediate rehabilitation efforts.

5.7.2 Guard Against Unscrupulous Ideas:

"If you are the master be sometimes blind, if you are the servant be sometimes deaf."

—R Buckminster Fuller

The overall tsunami rehabilitation process was being monitored by the LG of islands and also by a very senior officer. I will call him as Mr. X! He was also touring extensively to various islands including Katchal. During one of his visit to Katchal, he had seen the temporary jetty operating very effectively and enabling restoration of rehabilitation activities. It was brought to his notice that the draft alongside was only 3 meters. He thought that the rehabilitation woks were to be accelerated at a very high speed in this island for which bigger vessels should be brought to Katchal. That means, the approach channel leading to the temporary jetty from the entry point of Katchal till the berthing face of the temporary jetty was to be dredged and deepened. Out of blue, there came a suggestion from him that the channel leading to the temporary jetty at Katchal was to be dredged to obtain a depth of 7 meters urgently so that bigger vessels could come and berth there. Mr. Naresh Kumar and I were shocked when the news was conveyed to us.

He asked me, "Sekar, you are an expert in Ports and harbor. Tell me if this helps?"

Me: "Sir, Why can't we talk to the office of Mr. X and find out how this proposal emerged and with what purpose?"

He agreed and immediately instructed his Personal Assistant (PA) to connect to the office of Mr. X. When enquired, we were told that the decision had already been taken by them for dredging Katchal up to 7 meters and nothing more to be explained in this regard and we might proceed on action. Mr. Naresh Kumar was all the more upset by the way the reply came.

When the official communication of this proposal came, Mr. Naresh Kumar asked for my comments officially on file. My comments were:

"There is no doubt that the process of rehabilitation in Katchal can be accelerated if larger vessels could call on there.

If larger vessels are to call at Katchal, it is possible only if the approach is made deeper as required.

Assuming that the approach is deepened by dredging, the next pertinent question is where the bigger vessels are to go and berth.

There is no berthing structure in Katchal which is capable of taking berthing forces of bigger vessels.

The temporary jetty constructed is capable of taking smaller vessels of draft 2.5 meters only and this cannot be used by bigger vessels as the foundation or fixity of the structure cannot cater for larger vessels.

Moreover, the Kapanga temporary jetty is located at the river mouth and once the monsoon sets in, the water flow from the upstream to this location is going to be higher which increases the current in the location as well as lot of deposition of silt from upstream also is likely which will deposit over the dredged channel which would then warrant constant and continuous maintenance dredging.

Moreover, dredging is an activity which cannot be taken up without sub soil data as dredging companies require the same to decide on the type of cutting tool they may have to engage at the location. Without conducting sub soil investigation, it may not be practical for any agency to organize dredging.

Apart from these technical aspects, dredging is an activity which requires environment clearance from Ministry of Environment and Forest (MOEF). For approaching MOEF for clearance, we may have to conduct

Environmental Impact Assessment (EIA) Studies by observing the site conditions for a period of one year which is mandatory.

Keeping the above points in mind a suitable decision may kindly be taken please."

Mr. Naresh Kumar was thrilled to see my comments. Immediately he called me on telephone, "Sekar. Saw your comments. I am forwarding the same to the office of Mr. X. Ok?"

I agreed.

Next day, I got a call from Office of Mr. X calling me to their office for discussions. I informed Mr. Naresh Kumar and went there. I was taken to the officer.

He told me, "Mr. Sekar, the main idea of my proposal is to get the thing done fast. Why don't you people understand it? Why do you object to something which is good for the people?"

Me: "I have never objected to your proposals Sir. It is my duty to bring out all pros and cons to you so that you are aware of all facts and circumstances before you take a decision or I may fail in my duty to appraise you of the ground situations."

Then after returning to my office I had apprised Mr. Naresh Kumar of what had transpired in the office of Mr. X

In the next two days, I got a call from the Secretary to Government of India, Ministry of Sipping.

"Sekar, what is the problem in deepening the approach channel to jetty at Katchal?"

I understood what is happening and I explained all facts of pros and cons as I had written on the file earlier. The Secretary told again, "The defense people are very much annoyed on you. Why can't you at least call tenders for dredging? Let us see other things later."

I agreed and accordingly, floated short tenders by sending tender papers to all known dredging companies by collecting the details from other port trusts and also from websites seeking response within a week's time. As expected, many of the firms called for sub soil data from the location. One firm from Gujarat, without asking any questions, gave its quote.

Mobilization charges : Rs.25000000

Demobilization charges : Rs.25000000

Daily charges : Rs.2500000

Irrespective of the quantity dredged from the day the dredger starts from its origin till the day the dredger reaches its originating place from Katchal.

I was about to fall flat on the ground to see such an unethical quote. On receipt of the quote, I contacted Secretary of Ministry and briefed him of what is happening. He could also understand what was wrong in that. He told me to hold till he called back.

Within half an hour, Secretary from Ministry called me on telephone and said," Sekar, the office of Mr. X had escalated the matter to the Cabinet Secretary in the Prime Minister's office and I have been asked to go to Cabinet Secretary the day after tomorrow to discuss this matter. You do one thing. Prepare a power point presentation and come to Delhi tomorrow itself so that we can finalize our course of discussion tomorrow itself. I gladly accepted.

Next day I reached Delhi and met our Secretary in the Ministry with my power point presentation. He was happy at the matters presented. He also mentioned, "Sekar, as most of the reasoning are based on technical aspects, I would suggest you to attend the meeting with me and go ahead with the presentation to Cabinet Secretary after my initial remarks

The next morning, I went along with my Secretary to Prime Minister's office for the meeting where the officer from office of Mr. X was already sitting. When I wished him, he was shocked to see me there. The meeting went on well. The Cabinet Secretary gave a patient hearing. At the end of the discussions, he addressed to the officer from office of Mr. X and said, "I think we should not rush into this proposal with so many ifs and buts which may lead all of us into problems tomorrow. Let us wait and see and consider the same later if required."

We disbursed after the meeting. Outside the conference hall, the furious officer gave a scornful look at me and said, 'You fellows do not understand the importance of dredging at Katchal and came here all the way to block it." I did not respond. My Secretary from Ministry smiled at me meaningfully.

On reaching back at Port Blair, Mr. Naresh Kumar was eager to know what exactly happened in the meeting with the Cabinet Secretary in Delhi. I explained the whole thing to him without missing the dramatic feelings etc. After a while, he spoke to me, "Sekar, Can you find out who owns the dredging company who quoted for this from Gujarat and the details of the top guys in the company."

Coming back to office, I browsed through their website, collected all the required information. A surprise was waiting for us. I took the details collected to Mr. Naresh Kumar as I could not wait any longer. On seeing the details, his face also brightened with a big grin. He said, "My hunch is proved correct."

Yes. The surname of the Managing Director of the dredging company was exactly the same as that of the officer Mr. X! We stopped discussing about it thereafter! We had the satisfaction of stopping draining of the public money and of course and to face a big enquiry coming up on a later date due to the greediness of somebody.

5.7.3 Tsunami Jetty, Hut Bay- Emergency Met:

"The conventional definition of management is getting work done through people, but real management is developing people through work."

— *Agha Hasan Abedi*

Little Andaman was another island that was heavily hit during the tsunami and suffered major damages. The harbor of Little Andaman is located in Hut Bay. This harbor has rubble mound break water (the break water was constructed using large size granite boulders of uneven size and shape and raised as a mound and hence the name Rubble Mound) for a length of 1200 meters in a hockey stick pattern which protected the lee side of the harbor against high swells of waves. There was a deep water wharf on the lee side of the break water and a piled jetty too. The tsunami waves had eaten away parts of the jetty which had to be abandoned as it was already old. The break water was damaged partially at several stretches which needed replenishment. Moreover, the land mass had undergone subsidence ranging from 90 cm to 1.50metres between Port Blair and Great Nicobar, the southernmost island of the group. The land mass had risen to a tune of about 40 cm on the Northern parts of Andaman Islands in places like

Diglipur. This effectively affected the draft conditions for shipping. The lesser draft by 40 cm posed problems for some of the vessels as the depth available have become insufficient for them to berth at the harbors whereas the increase in depth by about 1.05meters in Camp Bell Bay harbor in Great Nicobar island resulted in smaller head way for vessels during berthing as the top level of the berthing wharf have come down by about 1.50 meters due to subsidence of the land mass.

Coming back to the situation at Hut Bay, the connecting bridge portion between wharf and the break water got damaged and so the wharf could not be used by ships unless the approach to break water is restored. In the absence of jetty as well as the wharf, Little Andaman was crippled in the rehabilitation efforts. The continuing aftershocks were also being experienced in the form of medium to light tremors and all the staff at our circle at Little Andaman was living in the open space in front of our Model House at Hut Bay which is located on a high terrain. Most of the department quarters were inundated by the tsunami waves which rose up to the first floor of all the buildings, i.e., 6 meters high. Because of this, there was a lot of mud and slush inside all the houses. This could not be washed and cleared as the water supply pipeline system was also badly damaged. Amidst this chaos, all the coordination required for sound leadership to sort out the issues. My wife who was very keen to put in her piece of service at this demanding situation, offered herself to serve at Little Andaman. We traveled by a cargo sailing vessel which navigates through sail and a single engine. Very brave lady! After loading required rehabilitation materials for works, some quantity of ration items, vegetables etc., we set sail for Hut Bay harbor by 7 pm from Port Blair.

The sea conditions were very rough. Given there were continuing aftershocks and tremors, there were some who feared another tsunami. In spite of the dangers, we were determined to march on. AS soon as we set sail, the sea water was splashing on us from all sides due to the pitching and rolling of the vessel. To keep ourselves enthusiastic, we were sitting on a bench in front of the wheel house at the top and singing at the highest pitch possible to prove to the adverse conditions that we cannot be cowed down!

Next morning at around 8 am, we reached Hut Bay. It was one of the longest nights of our lives!

We went straight to Model House, met all our staff and families. My wife took charge of all the ladies and engaged them in cleaning the area around and organized community cooking for all of us. She formed some teams for cooking, cleaning, water supply to people there, bringing vegetables and cutting it etc. I formed some teams for men for various tasks like restoration of water pipeline and restore water supply which was on top most priority, cleaning of all the residences for putting the people back into their houses, restoration of office equipment, tracing out all the files which had been washed away in water and strewn around a radius of 2 km, with some files being restored from top of trees too. Those files thus restored were spread on concrete surface for drying in the hot sun like drying papads! Slowly, the situation came under control over a period of a couple of days.

The next important task was to organize a berthing place for at least smaller vessels like vehicle ferries for transporting rehabilitation materials into the island. So, a berthing place where a minimum of 2.5 meters depth of water was required with an undisturbed free approach for such vessels. A location was so selected, old steel trays which were available in the quarry area of Little Andaman were collected and making use of the same, a temporary berthing head was created on war-footing basis. I owe the success of this project to Mr. Mohammed, Junior Engineer (Civil) whose innovation and ideas were implemented in this with his own personal supervision. This temporary jetty at Hut Bay was used extensively which was fondly referred to as Tsunami Jetty by the local population.

5.7.4 Shoot at the Air and Reap the Fruit:

"The secret of managing is to keep the guys who hate you away from the guys who are undecided."

— *Casey Stengel*

While the tsunami rehabilitation activities were in full swing all over the places, it had some problems in one particular island, not because of lack of facility, not because of lack of man power, not because of lack of funds, not because of lack of technical snag. One of the senior officer was behaving in a funny way - non cooperative because the person wanted a transfer to Port Blair which could not be considered favorably due to many reasons. But the officer was adamant not to understand the reasons and was trying to adopt unscrupulous methods to sabotage the projects and the quality of the deliverables.

When I came to know about this, I displayed strict methods. I immediately called for explanations and justification. The response I received was evasive and was mentioned that it was the responsibility of the other junior officer working under that person and so cannot take responsibility for it. I made sure that the person was not spared and made that person responsible officially.

When this tussle was going on, I got an anonymous letter by post addressed to me by name. There was no from address too on the cover. Then I started scrutinizing the cover on which I could find the seal of Rangat as the station of origin of the cover. The content of the anonymous letter was that I was harassing an officer at that particular island where the rehabilitations projects were getting stalled and that it was not correct as that person too belonged to the same state where I hail from and that I should have had the courtesy and a soft corner. That letter was written to me based on the request of that officer, it said.

It was so silly that somebody was writing an anonymous letter and also admitting that he was writing the letter based on the request of the person affected.

I started thinking as how a letter could emerge from Rangat while the person who wanted relief was in a different island. Then, I called my Office Superintendent to find out and tell me if anyone was recently got transferred from that island to Rangat. Soon he came up with a name who was a Junior Engineer. I knew him very well as he worked with me on various occasions. My hitch was that he was the person who wrote the letter.

I told my Personal Assistant to connect me on telephone to that JE at Rangat immediately. Once he was on the line, I started speaking in a cool voice," Hello Mr.…. I got your letter regarding that matter!" The very next moment, he replied, "Yes Sir. I am very sorry Sir. I told that person several times that I will not write to CEA. But I was forced by him to write and was given specific directions to write that letter. Please don't mistake me sir."

I laughed to myself and hung up.

Think fast, analyze the situation with open mind, get to the root of the issue, apply your mind and solve the problem instantly without any delay. You will be a runaway success!

5.7.5 My Ever Best Management Story-Real One:

Don't read success stories. You will get only a message.
Read failure stories. You will get many ideas to success.

Dr.A.P.J.Abdul Kalam

It was almost two years after the dreadful tsunami created havoc in Andaman and Nicobar Islands. A senior member from Planning Commission of India was visiting A & N Islands to review the progress of rehabilitation efforts being executed by various departments in the islands. Various engineering departments like Andaman Lakshadweep Harbour Works (A.L.H.W.), Military Engineering Services (M.E.S.), Andaman Public Works Department (A.P.W.D.), and Central Public Works Department (C.P.W.D.) were the leading departments on the job. On her visit to Port Blair, the Hon'ble Member was escorted in a helicopter to the Southern Group of Islands for inspection where damages were severe and this requiring more concentrated rehabilitation efforts. She was accompanied by senior I.A.S. officials of A & N Administration including Secretary (Shipping). On their return to Port Blair in the evening, an emergency meeting was organized in the conference hall of Secretariat to review and discuss on the observations made by the Member. Heads of all major departments were commissioned for the meeting. Me as Head of A.L.H.W. entered the hall wherein most of the senior officers were already present. On seeing me, The Secretary (Shipping) rushed to me and told in an exasperated voice:

"Mr. Sekar, The Member is very much annoyed on A.L.H.W and you. You are likely to be on the firing line for today. Be prepared to face the situation."

Me: "For what Sir?"

He: "The construction of permanent shelters is lagging behind so much and you are blamed for the delay."

Me: "But the construction of permanent shelters are being executed by C.P.W.D. and not A.L.H.W.?"

He: "I know that Mr. Sekar! C.P.W.D officials have complained to her that their progress was slow because A.L.H.W have not constructed berthing jetties at those locations, i.e., Pulobhabi, Pulokunji, Little Nicobar etc. and due to non-availability of jetties at these locations, their progress is extremely hampered. Member is very much upset with A.L.H.W."

Me: (Smilingly) "Nothing to worry Sir, I will handle this! Thank you for sharing this with me."

The meeting started at 5 pm and the member started sharing her observations on the rehabilitation efforts. There were various directions given to A & N Administration and other departments and organizations present in the meeting. Then the eventful item of construction of permanent shelters came for discussion. Again a senior officer while explaining for the delay was attributing their delay to non-availability of jetties in those locations. Madam spoke:

"Is there any responsible officer from A.L.H.W in this permanent shelter hall?"

Me: "Good Evening Ma'm! I am Sekar, Chief Engineer and Administrator of A.L.H.W."

Member: "Mr.Sekar, wherever I went, they are saying that you have not constructed jetty and that is why their progress is hampered. You are such a key department dealing with sensitive projects of port and harbor. Can you share with me why your departmentcouldn't stand up to the occasion of emergency?"

Me: "Ma'm, Tsunami rehabilitation projects (T.R.P.) for Ports and Harbors were formulated very systematically with several rounds of discussions with Port department, Secretary (Shipping) of A&N administration, Secretary (Shipping) Union Ministry of Shipping and also Planning Commission of India before the list was finalized. I am sure that you are aware of the fact that the T.R.P. projects were categorized as Short Term, Medium Term and Long Term and the documents were also wetted through Planning Commission.

During the past two years, we have completed almost all projects listed in short term and most of the projects listed in the Medium Term are in progress and we are meeting the targets in most of the cases.

Coming to the contentious issue of construction of jetties at the said locations, they are new constructions listed in Long Term. These are the locations that did not have jetties even before tsunami. Mam, you are well aware of the fact that any new port project requires extensive surveys and investigations like hydrographic survey, topographic survey, collection of wind, wave, current data (site specific), sub soil investigation to decide on the type of foundation etc. which takes at least one year as so much data is to be collected from site which we have been doing.

Apart from these surveys, Environmental Impact Assessment (E.I.A.) studies through experts in the field are also to be conducted with at least one year site specific data. Once E.I.A studies are completed, the final report shall be submitted to local government for clearance and with that clearance, we have to apply to Ministry of Environment and Forest for obtaining final clearance of the project. Only after this, the project is accorded Administrative Approval and Expenditure Sanction (A.A. & E.S.). Only then the process for tendering and execution starts. The required studies are in progress and as far as A.L.H.W. is concerned we are rightly placed as planned in our T.R.P. program.

It is surprising to hear that C.P.W.D.is now surprised to note that there are no jetties at these locations at this last moment and is blaming any delays on non-availability of jetty. This has never been discussed with me at any point of time.

Moreover, C.P.W.D. is a major department throughout the country and they are very much aware of the procedures involved while they formulate any project. So, before they could frame the estimate for these works of construction of permanent shelters at these locations, I presume that their site engineers must have visited these sites and understood the ground realities and understood that there are no jetties in these locations. So, we can presume that they must have factored this in their estimates.

While they called tenders for this work, they also must have briefed the contract agencies regarding the nature of works and the site conditions. There is a condition incorporated in the tender documents of all C.P.W.D.

tenders that the agencies before quoting their rates shall visit the site and get acclimatized with the ground situation and that no claim whatsoever shall be entertained on this account at any later stage. Any agency quoting for the tender is presumed that they have visited the site of work and are well aware of the ground situation as per tender conditions. So, we can safely presume that the contract agencies executing these jobs there are well aware of the fact that there exist no jetties.

When the department of C.P.W.D. and the contractors of C.P.W.D knew very well even before finalizing tenders, it is not clear how they suddenly found out now that there are no jetties and that is why they couldn't progress? Perhaps their clarification on this matter would help us better understand their situation so we can help them!"

As the Member had understood the situation very well she became so ferocious but diplomatically said to the officers of C.P.W.D. "You may please sit with Chief Secretary and set things straight. I don't want any extra claims raised for funds by C.P.W.D." Then, she turned to me and said, "But Mr.Sekar, the problem still persists. What do you suggest to accelerate this?"

Me: "Ma'm, when we construct a jetty at any new location, the situation is same as what they are facing. We never had any jetty arrangements for transporting and unloading our construction materials. We invariably carry our materials for construction by ships, unload them in the anchorage on barges, tow the barges to beach area nearby, unload materials on shore manually and start our construction. They should also do the same."

C.P.W.D. Officer: "Ma'm, we don't have barges. If A.L.H.W. can spare few barges we will do."

Me: "In fact, we ourselves need more barges to meet our work load. As additional barges are not available for us, somehow we are managing with available resources. We don't have facility to spare any to them."

Member: "It is frustrating. What then is the solution?"

Me: "Probably, a temporary bund can be constructed connecting shore to a depth of 2-3 meters into the sea where smaller crafts with materials can come alongside and unload, mam."

Member: "Sounds sensible. Can C.P.W.D. Do it?'

C.P.W.D.: "As this was not envisaged in our original estimate sanctioned, we can't do it as funds are not provided for this in it."

Member: "A perfect Govt. department's answer!" "Can Administration execute this bund with available funds?"

Chief Secretary: "Can C.P.W.D. execute it if we provide additional funds?"

C.P.W.D. "I will discuss with my Chief Engineer and revert back to you Sir.'

Member:" Please understand the urgency. You should proactively come up to do it. I am sorry the way C.P.W.D. is handling this. Ok Mr. Sekar, can u do it?"

Me: "We would love to do mam. But, we are already running short of technical staff acutely in our department and already I have requested permission of our Ministry to fill up about 50 Junior Engineers. But A.P.W.D. is another engineering department of Administration and they are also capable of doing this mam.'

On hearing this, the Chief Engineer of A.P.W.D. threw a scornful look at me.

A.P.W.D.: "Ma'm, even for that, huge quantities of stone boulders are to be transported before formation of the bund. The quarries too do not have so much of products to cater even for the requirements of construction. There is acute shortage of quarry products mam."

Member: "Back to square one! Is there anything we can do?"

Me: "Yes mam. There is no need for any boulders or quarry materials from quarry for the bund!"

Member: "'Come on Mr. Sekar! Tell me!"

Me: "What is needed is to collect all empty cement bags from all nearby sites. Transport them in small fisherman boat to location and unload at site. Fill them with the beach materials from the shore and tie them securely. Stack them neatly one over the other for a width of say 2 meters and for the required length till sufficient water depth for berthing of smaller vessels is obtained. We can add some tyres at the face of the bund as fenders to avoid damage to the bund. Of course, as this is a temporary bund, this requires replenishment in case of any breach. And this can be done easily by A.P.W.D. Only some labour force is required."

Member: "There cannot be any simpler solution than this. So, A.P.W.D. will do this on war-footing-basis."

Me: "Ma'm, let me also share with the group that as this involves construction of the bund in the C.R.Z Zone, (Coastal Regulation Zone) we may probably need clearances from Environmental angle. As Principal Chief Conservator of Forests (P.C.C.F.) is also here with us, he may throw some light on this issue."

P.C.C.F: "Yes Mam. This attracts local body clearance."

Member: "My God! How fast can it be done?"

P.C.C.F.:"As soon as A.P.W.D. submits application in the prescribed format, we will arrange a meeting of the local body to clear it mam."

Deadlines were fixed for A.P.W.D., and P.C.C.F and the marathon meeting ended finally. After the exit of the Member from the hall, the Secretary (Shipping) who sounded an alarm before the meeting came to me and congratulated me saying "'Kudos! You not only made it but pulled every one up effectively and rightly so. Great going! Keep it up!"

Soon after the Member returned back to Delhi, she had a meeting with Senior Officials of Ministry of Shipping after which I got clearance for filling up of 51 vacant posts of Junior Engineers on an immediate basis!

I very often cherish this memory of this incident, particularly because my colleagues and my office were extremely happy and appreciative of this. I took pride in the courage I summoned in placing facts without fear, which is normally lacking in the Government services as most people wouldn't dare to speak in front of very senior officers, especially from Central Ministry. My staff was also amused to hear how I pulled up all departments while emerging clean from the expected onslaught on our department!

I always tell my people, "Be confident! Get to know yourself and grow with new experiences. Don't be held back by a "fear". Strive for your goals." I knew I was practicing it! So I never hesitated to tell this to my people. At every given opportunity of addressing my people, I used to insist that they should never compromise on Quality and Time in any work they do and they should feel that nobody in the world could have done it better than them in the true sense.

Don't let people knock you down. If anything, let them get knocked down to prove you can be much better than them.

This had encouraged my staff to come out for open discussions with me on any subject to improve efficiency of working. Once when I was telling my people that they should always strive hard to follow the codes and manuals of works and accounts, one of my Engineer asked me,

"Sir, How can we read and remember the codes and manuals? Do you want us to spend our time in reading only or to work on the projects?"

The question is very relevant to most of them as they were nodding in acknowledgement of the same.

I said, "The codes and manuals are meant for right execution of projects in the righteous way only. So, anything done in contrary to them is objectionable and we are answerable for it. Simple suggestion from me is this. Whatever you do in any project, check with your conscience. If there is even a slightest objection from your conscience, please go and refer to the relevant chapters of the code or manual. When there is no irking from your conscience, those things find a clearance from Manual and codes and when you have even slightest doubt from your conscience don't do it unless you get it referred through manual and verified for its correctness."

The Executive Engineer in the Office of Chief Engineer and Administrator and almost my Technical P.A. was Mr.Alok Srivastav once asked me when I took over the reigns of CEA, what I expected from his technical section as for the paper works were concerned. Probably, he might have thought that I might not like if something against my ideas were written on the file. I told him point blank that he should not hesitate to point out if something was going wrong and even if I directed him to do something which might not fall under permitted code or ethics, he should not do it and he should write his views clearly and fearlessly on the files. Right should be right and wrong should be wrong always. He should also read the guidelines issued by Central Vigilance Commission from time to time and ensure that every direction is strictly followed so that we as the apex office should show other subordinate offices what they should do. These were taken as words of Bible and followed strictly.

6.0 Epilogue

Learning is a never-ending process and no one in the world has mastered any field despite many credentials such as 'Masters'. Even people that hold Masters and Doctorates will surely agree that their education is continuous till date are that there's still much to learn and assimilate. We learn from books, teachers, personal experiences and from other's success as well as failures too. My learning strictly based on books or school education in Management is restricted to a Diploma in Management through distance education and that was done casually propelled by an interest to diversify into various realms of learning opportunities as I grew up the ladder in government service. So I can't say that I have mastered the theories in Management. But I am happy to find that I possessed the natural instinct towards it and that immensely helped my career and equipped me to learn through experiences as was the demand at my position. I have also come across many people, who are applying management principles in their day to day operations, without formal education in the field. It is my sincere hope that this book helps those professionals find satisfactory answers to the questions in their minds and are able to successfully bridge the gap between their professional behaviors and leadership traits.

So I strongly believe that though higher education in Management is of course a necessity for making a manager the best suited to his profession, he or she becomes a matured Manager only through experience which gives lot of confidence to deal with situation. The different concepts of management are understood more clearly when practical examples prove it affirmative.

During a robbery in a bank, the robber shouted to everyone in the bank, "Don't move. The money belongs to the government. Your life belongs to you."

Everyone in the bank lay down quietly.

This is called "Mind Changing Concept", changing the conventional way of thinking.

When a lady started creating a raucous, the robber shout back, "Please be civilized. This is a robbery and not a rape."

This is called "Being Professional." Focus only on what you are trained to do!

When the bank robbers returned home, the younger robber(MBA trained) told the older robber(who has only completed year 6 in primary school), "Big brother, let us count how much we got."

The older robber rebutted and said, "You are very stupid. There is so much money and it will take us a long time to count. Tonight, the TV news will tell us how much we robbed from the bank!"

This is called "Experience." Nowadays, experience is more important than paper qualifications!

After the robbers had left, the bank manager told the bank supervisor to call the police quickly. But the supervisor said to him, "Wait! Let us take out $ 10 Million from the bank for ourselves and add it to the $70 Million that we have previously embezzled from the bank."

This is called "Swim with the Tide." – Converting an unfavorable situation to your advantage.

The supervisor says, "It will be good if there is a robbery every month." This is called "Changing Priority."

Next day, the TV news reported that $100 Million was taken from the bank. The robbers counted and counted, but they could count only $20 Million as their booty. They were very angry and started fuming, "We risked our lives and could take only $20 Million. The bank manager looted $80 Million with a snap of his fingers. It looks like it is better to be educated than to be a thief!"

This is called "Knowledge is worth as much as Gold!"

It is an undisputable fact that education complemented with experience and applying the knowledge gained in a most prudent manner will go a long way in most productive and positive management in any field one chose to be.

My experiences which were my experiments gave me lots of lessons, anguish, surprises, pain, anger and of course joy, happiness and enormous satisfaction at the end. The end justified my means irrespective of the correctness or otherwise of the techniques adopted in dealing with the cases.

The most important aspect of success is attitude and determination to fight back in a prudent and acceptable manner and adapting to the prevailing conditions so that one is not deprived of success.

To be most successful, a Novice shall show his utmost sincerity and perseverance to understand situation and learn with all eagerness. A middle level manager shall strive hard to apply his or her mind to improvise and show new propositions to suit the situations and be able to convince his or her boss and gain fullest confidence of the boss who would think that you are an asset to your organization. An Executive boss shall keep his or her cool at testing times, guide subordinates out of their problems and prove himself or herself as a real boss to the appreciation of his team. He or she shall display highest degree of integrity to the organization. Top Executive shall have overall responsibility to understand the staff at each level, their problems if any, their bottle necks, requirements and try to provide the best environs for their working to get the best results. Top Level Executives shall have the maximum humility and they shall not be terrorizing the staff in the organization. Once this level of maturity is reached at each level, then there is no stoppage for an excellent growth of the organization.

I wish everyone a grand success in all their endeavors and best of luck!

References

1) Pic. Dry dock at Port Blair Source: http://alhw.and.nic.in/Drydock. htm. Section 2.1

2) Pic. Old Type-Timber Pile Driving Frame. Source: http://alhw.and. nic.in/ Section 2.1.1

3) Pic. Multiple pile driving in Andaman Islands. Source: http://alhw. and.nic.in/ Section 2.1.3

4) ***"Success happens not by chance......*** should always look to first is yourself." Source: www.pickthebrain.com, URL: http://www. pickthebrain.com/blog/7-undeniable-reasons-why-some-people-fail-where-others-succeed/ Section 2.2.1

5) "There's always more to know........ what they've learned with others."Source: http://www.facultyfocus.com/articles/teaching-professor-blog/seven- characteristics-good-learners/ Section 2.2.1

6) "Management utilizes all the inputs....of any resources." Source: http:// www.managementstudyguide.com/management importance.htm. Section 2.2.2

7) "Long ago, there was a battle....... great victories will be won." Source: http://betterlifecoachingblog.com/2012/04/27/the-wise-general-a-story-about-the-value-of-self-belief by Darren Poke Section 2.2.3

8) "According to research at Najafi Global Mindset Institute....... smoother and less stressful." Source: www.globalmindset.com. Section 2.2.4

9) "IT'S HARD TO BE CONFIDENT IN YOURSELF...... MAGAZINES, AND EDUCATIONAL INSTITUTIONS." SOURCE: HTTP://ZENHABITS.NET/25-KILLER-ACTIONS-TO-BOOST-YOUR-SELF-CONFIDENCE/ SECTION 2.2.4

10) "One businessman was in debt and couldn't figure out a way out of it......, to achieve everything that he had now." Source: https://www. lifepositive.com/leap-faith/ under title: **Leap of Faith** Section 2.2.4

11) ""*It is a law of nature we overlook*....... says <u>H.G. Wells</u> in his "<u>The Time Machine</u>". Source: H. G. Wells' *The Time Machine* was produced by <u>John Walker</u> in May 2002.Section 2.2.5.

12) "This is the inspiring story of a furniture store.... sure to be told for years to come." Source: <u>http://www.printingforless.com/blog/ custom-printing/adaptability-a-real-life-success-story</u> Section2.2.5

13) "Create an atmosphere that is open...... organization for more support." Source:Article by Chapman,V <u>http://usemyability.com/ resources/skills_abilities/flexibility-and-adaptability.html</u> Section 2.2.5

14) "Two cockroaches lived in a house.......saved the life of the cockroach."Source: <u>http://shortstoriesshort.com/story/believe-in- yourself/</u> Section 2.2.6

15) "In order to turn the mind toward the positive......... and can be used together with it." Source: http://www.successconsciousness. com/index_000009.htm.Section 2.2.6

16) "Self discipline is not a new idea....... *your backbone should be!*" *Source: Author:* Mark Tyrrell. *Title* "Self discipline and mental health" in page <u>http://www.uncommon-knowledge.co.uk/self- discipline.html Section 2.2.7</u>

17) "A man in a hot air balloon realized he was lost...... somehow, it's my fault." *Source:* Page: <u>http://creativebits.org/the_man_in_the_hot_air- balloon Section 3.0</u>

18) "A teacher and his student were walking.... concluded the teacher." Source: by Ramez Sasson insuccessconsciousness.com Section 2.2.7

19) "Failing to plan is planning to fail....... see true meaning in your work" Source: 10 ways to improve your performance at work *by <u>Lama Ataya</u>, June 5, 2013 Page: <u>http://www.wamda.com/2013/06/10-ways- to-improve-your-performance-at-work</u> Section 3.1*

20) "A lot of research has been carried out on this theory.......... And so is the case with business." Source: <u>http://www.brainyquote.com/ quotes/authors/r/roger_staubach.html</u>. Section 3.2.

21) "Business growth requires a track record of success.......... same to fellow teammates or even customers." Source: http://www.inc.com/ eric-v-holtzclaw/consistency-power-success-rules.html Section 3.2

22) "When our inner systems (beliefs, attitudes, values, etc.)........ inner systems and social norms." Source:<u>http://changingminds.org/ explanations/theories/consistency_theory.htm</u> Section 3.2

23) "Once there were two friends........... He left putting in the effort till too late."*Source:<u>http://www.experiential-learning_games.com/ leadershipconsistency.html.</u>Section3.2*

24) "If you were to test for analytical skills you might be asked....... problems with their analytical mindset."Source: Analytical Mindset. Page: <u>https://www.boundless.com/business/textbooks/ boundless-business-textbook/management-8/characteristics-of-good-managers-63/analytical-mindset-304-7874/</u> Section 3.3

25) "The modern mind tends to be more…That analytical background really helps."*Source:* BrainyQuote.<u>http://www.brainyquote.com/ inquire/inquire.html#HiZcuk5e2yShstEp.99</u>) Section 3.3

26) "With the global nature of our business….. articulated, and respected."Page: "Building Effective Business Relationships". Page: <u>Leading Insight, Orange County California</u> http://www. leadinginsight.com/ Section 3.4.

27) "One of the interesting things that we see in business meetings….. no one will want to talk to." Author: Neil Fogarty Page: *<u>http:// www.virgin.com/entrepreneur/8-tips-relationship-building-business</u>* Section 3.4

28) "Rama *Rajya* is the benchmark of good governance...... relevant today as it was in theTreta period."Source: Titled: Ramayana's **vision has eternal relevance;** by *<u>Pramod Pathak</u> in URL: <u>http://dailypioneer. iws.in/columnists/spirituality/the-essentials-of-good-governance.html</u> Section 3.4*

29) *Pic of*Slipway at Port Blair. Source:<u>http://alhw.and.nic.in/SLIPWAY. htm.</u>*Section 3.5.6*

30) Pic of Austin Bridge at Mayabunder Source:<u>http://alhw.and.nic.in/ Roads%20&%20Bridge.htm</u>. Section 3.5.7

31) "Three stone-cutters were engaged in erecting a temple........ vision in one's work for the common good." Source: Boloji.com. Author: M. P. Bhattathiry in page <u>http://www.boloji.com/index.cf m?md=Mobile&sd=Articles&ArticleID=1180</u> Section 4.1

32) "Reaching the end of a job interview….. but you started it."Source: http://www.laughfactory.com/jokes/office-jokes *"Used with permission from Jokerz Media, LLC"* Section 4.1

33) *"A project manager, hardware engineer and software engineer………* *and see if it happens again". Source* page http://www.cvr-it.com/PM_Jokes.htm Section 4.2

34) "Engineers and scientists are generally seen as never making……. The less you know, the more you make." Source: kate piersanti www.bkconnection.com. Section 4.2

35) "The concept of deliberation in the present use…….. change actors' attitudes or beliefs if they can be"Source: Eriksen, Erik O. (2013) 'Reason-Based Decision-Making: On Deliberation and the Problem of Indeterminacy', ARENA Working Paper 06/2013, Oslo: ARENA Centre for European Studies, University of Oslo. Available at: <http://www.sv.uio.no/arena/english/research/publications/arenapublications/workingpapers/working-papers2013/wp6-13.pdf> Section 4.4

36) "Following is a short story by Pedro Pablo……. I will turn this little job into something really great". Source: http://freestoriesforkids.com/children/stories-and-tales/insignificant-task. Section 4.6

37) "A NEW MANAGER SPENDS A WEEK AT HIS NEW OFFICE….."PREPARE THREE ENVELOPES." SOURCE: THREE ENVELOPES. BY MYINT ZU WIN HTTPS://WWW.LINKEDIN.COM/PULSE/THREE-ENVELOPES-MYINT-ZU-WIN SECTION 4.7

38) "Stever Robbins is president of VentureCoach.com, a Cambridge,…… they don't know how to help it get there."Source: https://hbr.org/2009/03/seven-communication-mistakes-m.html Section 5.5

39) "If we only take the easy route…… You can do it!" Source: Method 2 of 3: Furthering Good Habits. Page: http://www.wikihow.com/Believe- in-Yourself Section 5.6

40) "Successful leaders recognize the need to adapt….. can lose business to the opposing team." Source: Under heading "How to Get the Competitive Edge?" Page: http://study.com/academy/lesson/

what-is-global-competition-in-business-definition- challenges-quiz.
html Section 5.6

41) "Motivated by the desire for achievement...... stability over risk, intimacy, and self-discovery."Source: : article under title of book "MARKETING STRATEGY AND MANAGEMENT BY MICHAEL J. BAKER"INURL:HTTPS://BOOKS.GOOGLE. CO.IN/BOOKS?ID=W69HBQAAQBAJ&PG=PA273&LP G=PA273&DQ=MOTIVATED+BY+THE+DESIRE+FOR+ ACHIEVEMENT,+ACHIEVERS+HAVE+GOALORIENT ED+LIFESTYLES+AND+A+DEEP+COMMITMENT+TO +CAREER+AND+FAMILY.&SOURCE=BL&OTS=DENG WPLKGL&SIG=LJJKRPWBHYAMRNMZ5OAKUWIB WXK&HL=EN&SA=X&VED=0CCCQ6AEWAMOVCH MIR4H2JMQBYAIVWZ6OCH0SRG6Y#V=ONEPAGE& Q=MOTIVATED%20BY%20THE%20DESIRE%20FOR%20 ACHIEVEMENT%2C%20ACHIEVERS%20HAVE%20 GOAL-ORIENTED%20LIFESTYLES%20AND%20A%20 DEEP%20COMMITMENT%20TO%20CAREER%20AND%- 20FAMILY.&F=FALSE Section 5.6

42) "Start with a cage containing five monkeys... how company policies are made." by Michael Michalko from his book Thinkertoys. Section 5.2

www.ingramcontent.com/pod-product-compliance
Lightning Source LLC
Chambersburg PA
CBHW030006190526
45157CB00014B/445